Iterative Solution of Symmetric Quasi-Definite Linear Systems

SIAM Spotlights

SIAM Spotlights is a new book series that comprises brief and enlightening books on timely topics in applied and computational mathematics and scientific computing. The books, spanning 125 pages or less, will be produced on an accelerated schedule and will be attractively priced.

Josef Málek and Zdeněk Strakoš, *Preconditioning and the Conjugate Gradient Method in the Context of Solving PDEs*

Paul G. Constantine, *Active Subspaces: Emerging Ideas for Dimension Reduction in Parameter Studies*

Dominique Orban and Mario Arioli, *Iterative Solution of Symmetric Quasi-Definite Linear Systems*

Iterative Solution of Symmetric Quasi-Definite Linear Systems

Dominique Orban
GERAD and École Polytechnique
Montréal, Québec

Mario Arioli
Emory University
Atlanta, Georgia

siam.
Society for Industrial and Applied Mathematics
Philadelphia

10 9 8 7 6 5 4 3 2 1

Publisher	David Marshall
Acquisitions Editor	Elizabeth Greenspan
Developmental Editor	Gina Rinelli Harris
Managing Editor	Kelly Thomas
Production Editor	Louis R. Primus
Copy Editor	Bruce Owens
Production Manager	Donna Witzleben
Production Coordinator	Cally Shrader
Compositor	Scott Collins
Graphic Designer	Lois Sellers

Library of Congress Cataloging-in-Publication Data

Names: Orban, Dominique. | Arioli, M. (Mario)
Title: Iterative solution of syammetric quasi-definite linear systems / Dominique Orban, Mario Arioli.
Description: Philadelphia : Society for Industrial and Applied Mathematics, [2017] | Series: SIAM spotlights ; 3 | Includes bibliographical references and index.
Identifiers: LCCN 2016053335 (print) | LCCN 2016054562 (ebook) | ISBN 9781611974720 (print) | ISBN 9781611974737 (ebook)
Subjects: LCSH: System analysis. | Iterative methods (Mathematics) | Differential equations, Linear. | Linear systems.
Classification: LCC QA402 .O686 2017 (print) | LCC QA402 (ebook) | DDC 003/.74--dc23
LC record available at https://lccn.loc.gov/2016053335

To my mother and father,
who worked hard to make this possible.
-Dominique

To my wife, my family, and my friends.
-Mario

Contents

List of Algorithms

List of Theorems

Preface

The research that led to the work described here was initiated during the winter of 2010 by informal discussions at the Rutherford Appleton Laboratory in the beautiful countryside surrounding Oxford. That numerical methods for optimization and the simulation of partial-differential equations using the finite-element method need to solve systems with similar structure is nothing new. Until recently, the selection of iterative methods for symmetric and quasi-definite systems was probably reduced to MINRES and SYMMLQ, which do not exploit the structure beyond symmetry. Linear algebra is a beautiful unifying topic in computational science, one that can be discussed and developed across disciplines and fields of applications. We feel that that is exactly how the present research materialized.

Background

Defining the background necessary to read this book is challenging because the material draws from the bottomless literature on linear least-squares problems and Krylov methods. We assume knowledge of basic linear algebra at an advanced undergraduate or early graduate level. We also assume that the reader has encountered a linear least-squares problem before and is able to recognize the optimality conditions of convex quadratic problems. No additional knowledge of applications, such as optimization or fluid dynamics, is required, though it might help understand the context of the numerical experiments of Chapter 7. Familiarity with factorization-based methods for least-squares problems will help the reader understand the core of the iterative methods and put things in perspective but is not strictly necessary. Part of the challenge of understanding Krylov methods is realizing that each method solves a well-defined optimization problem at each iteration. We organized the discussion so that sections describing implementation details can be skipped or skimmed if desired. This will let the reader have an overview of methods and their purpose and leave implementation details for later. Prior familiarity with one iterative method based on the Lanczos or Golub–Kahan process will greatly help but is not strictly necessary either.

Audience

The material described in the present book is suitable for a graduate audience and researchers in computational science. The application areas with which we are most familiar are numerical linear algebra, computational optimization, and computational finite-element methods. However, we encounter new applications areas every day where symmetric and quasi-definite systems occur, including machine learning, geophysics, medical imaging and numerous other fields where inverse problems play a central role, weather forecasting, and many more. The book could be used as the central reference in an advanced graduate-level course on computational methods for least-squares problems or saddle-point systems. The material can also be selectively sampled to provide background on numerical methods in other fields. For example, Section 2.2, Chapter 4, and Sections 5.1 and 5.2 were used to provide concrete numerical methods used in a two-week accelerated course on factorization-free optimization.

Software

Computational scientists are often opinionated, and it can be difficult to convince them to use tools that are not already part of their tool set. One such tool is their favorite programming language. As it happens, we are enthusiastic about programming languages, and in an attempt to satisfy ourselves and the largest possible audience, we provide implementations in three programming languages to accompany and complement this book. Any knowledge of computational methods is incomplete and inaccurate until one got his or her hands dirty playing with an implementation, running it on test problems, examining the behavior of a method in finite-precision conditions, and understanding the results. Computational scientists rarely have the background necessary to produce the high-quality, high-usability, user-friendly software that computer scientists and computer engineers are trained to produce. We are no exception to the rule, but we did our best to make the software easy to access, easy to install, and easy to use. Time will tell if we succeeded. Readers will find implementations in MATLAB, Python, and Julia of all methods described in the book available from www.siam.org/books/sl03. We encourage all users to get in touch and let us know how they fare. We especially welcome reports letting us know what can be improved.

Acknowledgments

Mario Arioli wishes to express his gratitude to several institutions for inviting him to spend several sabbatical months in their mathematics and computer science departments: Emory University, which awarded him an Emerson fellowship and later appointed him as adjunct professor; the Berlin Mathematical School and the Technische Universität Berlin; the CIMI and the Paul Sabatier University (Toulouse); the INPT (Toulouse); and the Bergische Universität Wuppertal. He also thanks several friends for their valuable help and advice: Volker Mehrmann, Andreas Frommer, Iain Duff, and Daniel Ruiz.

Dominique Orban wishes to express his gratitude to the Rutherford Appleton Laboratory for hosting him for four sabbatical months in the Numerical Analysis Group, where this research was born. Work conducted while at Rutherford Appleton Laboratory was sponsored by EPSRC Grant EP/E053351/1. The remainder of the work was partially supported by NSERC Discovery Grant 299010-04. Much of the material here has greatly benefited from fruitful discussions with my good friend Michael Saunders. Indeed, iterative methods for linear systems and least-squares problems would not be where they are today without him.

We both also want to sincerely thank three referees who kindly provided their time as well as suggestions and advice on the text and the presentation.

Paige (1974) and Saunders (1995) have been constant sources of inspiration during the development of this research. We owe them much for paving the way ahead. We also wish to express our gratitude to Michele Benzi of Emory University for fruitful discussions.

Chapter 1
Introduction

The solution of linear systems of equations has always been and remains a fundamental subproblem of countless computational methods for numerical science and engineering. The method of choice to solve systems is typically guided by their properties, including symmetry, definiteness, sparsity, conditioning, and block structure. Numerous applications, including constrained optimization and fluid dynamics, involve systems with the specific *saddle-point* block structure; i.e., their coefficient has the form

$$\begin{bmatrix} \mathbf{F} & \mathbf{B}_2^{\mathsf{T}} \\ \mathbf{B}_1 & \mathbf{G} \end{bmatrix},$$

where the block $\mathbf{F}$ is square but need not be symmetric, the block $\mathbf{G}$ is also square and can be zero, and $\mathbf{B}_1$ may or may not be equal to $\mathbf{B}_2$. The special case of *symmetric saddle-point systems* arises when $\mathbf{F}$ is symmetric, $\mathbf{B}_1 = \mathbf{B}_2$, and $\mathbf{G}$ is either zero or symmetric and negative semidefinite. At the time of this writing, a large number of special cases of saddle-point systems, including the symmetric case, are understood sufficiently well that efficient numerical methods are known to solve them, though it is these authors' opinion that such systems still have much to reveal. The state-of-the-art paper by Benzi, Golub, and Liesen (2005) covers the main properties of general saddle-point systems and families of associated numerical methods. This book does not attempt to give a complete overview of saddle-point systems, of their properties, or of numerical methods, as they are the topic of ongoing research. Rather, we focus our efforts on a special case of symmetric saddle-point systems that arises when numerical methods are somehow *regularized*. Thus, we recommend that the interested reader or student who is not yet familiar with saddle-point systems begin with Sections 1–9 of Benzi, Golub, and Liesen (2005) before going further.

A central argument of this book is the connection between symmetric saddle-point systems and certain optimization problems. We can already state a simplified version of the key relationships to keep in mind at this point. Consider a full-rank tall and thin matrix $\mathbf{A}$, i.e., $\mathbf{A}$ possesses n rows and m columns with $n > m$, and let $\mathbf{I}$ be the $n \times n$ identity matrix. Then, for vectors $\mathbf{b}$ and $\mathbf{c}$ of appropriate size, the linear systems

$$\begin{bmatrix} \mathbf{I} & \mathbf{A} \\ \mathbf{A}^{\mathsf{T}} & \end{bmatrix} \begin{bmatrix} \mathbf{r} \\ \mathbf{x} \end{bmatrix} = \begin{bmatrix} \mathbf{b} \\ 0 \end{bmatrix} \quad \text{and} \quad \begin{bmatrix} \mathbf{I} & \mathbf{A} \\ \mathbf{A}^{\mathsf{T}} & \end{bmatrix} \begin{bmatrix} \mathbf{y} \\ \mathbf{z} \end{bmatrix} = \begin{bmatrix} 0 \\ \mathbf{c} \end{bmatrix} \tag{1.1}$$

are equivalent to the optimization problems

$$\underset{\mathbf{x}}{\text{minimize}}\ \tfrac{1}{2}\|\mathbf{A}\mathbf{x}-\mathbf{b}\|_2^2 \qquad \text{and} \qquad \underset{\mathbf{y}}{\text{minimize}}\ \tfrac{1}{2}\|\mathbf{y}\|_2^2 \ \text{ subject to } \mathbf{A}^T\mathbf{y}=\mathbf{c}, \tag{1.2}$$

respectively. The equivalence is in the sense that the systems (1.1) represent the necessary and sufficient optimality conditions of the optimization problems (1.2). We will refer to the first problem of (1.2) as the *least-squares* problem and to the second as the *least-norm* problem. The reader can verify that under the conditions given, the saddle-point systems above are nonsingular and the optimization problems are strictly convex. In the first problem, the vector $\mathbf{r} := \mathbf{b}-\mathbf{A}\mathbf{x}$ represents the least-squares residual. In the second problem, $\mathbf{z}$ is the vector of Lagrange multipliers associated to the equality constraints. As a minor point of terminology, we note that in optimization, symmetric saddle-point systems such as those above are sometimes called *augmented* systems or KKT systems. The terminology "KKT" is shorthand for "Karush–Kuhn–Tucker," which is the name given to the first-order optimality conditions of a smooth optimization problem, after their discoverers.

It will be important to go back and forth between saddle-point systems and least-squares or least-norm problems, but it will also be important to juggle with numerical methods for symmetric systems and numerical methods for least-squares and least-norm problems. For instance, the systems (1.1) are symmetric and indefinite, and if solving them with a direct method is of interest, one might suggest the sparse symmetric indefinite factorization. Once we realize that those systems arise from least-squares and least-norm problems, another possibility emerges. Both systems could be solved with a **QR** factorization of **A**. A third possibility arises when we reduce the saddle-point systems (1.1) to the form

$$\mathbf{A}^T\mathbf{A}\mathbf{x}=\mathbf{A}^T\mathbf{b} \qquad \text{and} \qquad \mathbf{A}^T\mathbf{A}\mathbf{z}=-\mathbf{c} \tag{1.3}$$

by elimination. Indeed, the matrix of (1.3) is symmetric and positive definite. Thus, a Cholesky decomposition could also be employed. In the text, we refer to (1.3) as the normal equations and the Schur-complement equations. The second system is also sometimes called the normal equations of the second kind. The connection between the **QR** factorization of **A** and the Cholesky decomposition of $\mathbf{A}^T\mathbf{A}$ is well known, namely, that $\mathbf{R}^T$ is the Cholesky factor of $\mathbf{A}^T\mathbf{A}$, and the reader should keep it in mind because it parallels a similar connection between iterative methods for (1.2) and for (1.3).

Broadly speaking, iterative methods that aim to solve (1.1) directly are based on the Lanczos (1952) process, which transforms a symmetric matrix to tridiagonal form. Such methods chiefly include MINRES and SYMMLQ of Paige and Saunders (1975), designed specifically for symmetric indefinite systems.

They also include the conjugate gradient method, denoted CG, of Hestenes and Stiefel (1952), though good behavior of the latter on indefinite systems is typically not guaranteed.

Because the systems (1.3) are symmetric and positive definite, CG is a natural solution method. There is a parallel with our discussion on factorizations because CG applied to (1.3) can be described as computing the Cholesky factorization of the tridiagonal generated from $\mathbf{A}^T\mathbf{A}$.

Methods for problems defined by rectangular matrices, such as (1.2), have a slightly different flavor and are typically based on the Golub and Kahan (1965) process, which transforms a rectangular matrix to upper (or lower) bidiagonal form. Successful methods based on a bidiagonalization include LSQR of Paige and Saunders (1982); the method of Craig (1955), which is usually denoted CRAIG; and LSMR of Fong and Saunders (2011). Each one of them can be described as computing a certain **QR** factorization of a bidiagonal matrix. In exact arithmetic,

- LSQR applied to the least-squares problem of (1.2) is equivalent to CG applied to the normal equations (1.3),
- CRAIG applied to the least-norm problem of (1.2) is equivalent to CG applied to the normal equations of the second kind (1.3),
- LSMR applied to the least-squares problem of (1.2) is equivalent to MINRES applied to the normal equations of (1.3).

In the same way that solving (1.3) using the **QR** factorization of **A** is usually more accurate than the Cholesky factorization of $\mathbf{A}^T\mathbf{A}$, it is also true that LSQR, CRAIG, and LSMR are usually more accurate than CG and MINRES applied to (1.3), especially on ill-conditioned systems.

As an exercise of generalization, consider the case where **A** is allowed to be rank deficient. The least-squares problem now possesses infinitely many solutions. It is no longer guaranteed that the least-norm problem is solvable because its constraints may be inconsistent. In this more delicate scenario, a common solution is to *regularize* both problems by introducing a small perturbation. The aim of this perturbation is to make the objective of the least-squares problem strictly convex again and to make the constraints of the least-norm problem consistent again. The regularized least-squares problem takes the form

$$\underset{\mathbf{x}}{\text{minimize}}\ \tfrac{1}{2}\|\mathbf{A}\mathbf{x}-\mathbf{b}\|_2^2+\tfrac{1}{2}\|\delta\mathbf{x}\|_2^2,$$

where $\delta \neq 0$ is a small constant. We are now solving a perturbed problem, but if $|\delta|$ is suitably small, we find comfort in the fact that we have introduced a small change into the problem to make it solvable again—see Björck (1996) for more information on the relationship between the regularized and original problem. The regularized least-norm problem takes the form

$$\underset{\mathbf{y},\mathbf{z}}{\text{minimize}}\ \tfrac{1}{2}\left(\|\mathbf{y}\|_2^2+\|\mathbf{w}\|_2^2\right) \quad \text{subject to } \mathbf{A}^T\mathbf{y}+\delta\mathbf{w}=\mathbf{c},$$

and we see that the two problems now involve

$$\begin{bmatrix}\mathbf{A}\\ \delta\mathbf{I}\end{bmatrix} \quad \text{and} \quad \begin{bmatrix}\mathbf{A}^T & \delta\mathbf{I}\end{bmatrix}.$$

Working backwards, we introduce $\mathbf{r}:=\mathbf{b}-\mathbf{A}\mathbf{x}$, and the systems corresponding to (1.1) now read

$$\begin{bmatrix}\mathbf{I} & \mathbf{A}\\ \mathbf{A}^T & -\delta^2\mathbf{I}\end{bmatrix}\begin{bmatrix}\mathbf{r}\\ \mathbf{x}\end{bmatrix}=\begin{bmatrix}\mathbf{b}\\ 0\end{bmatrix} \quad \text{and} \quad \begin{bmatrix}\mathbf{I} & \mathbf{A}\\ \mathbf{A}^T & -\delta^2\mathbf{I}\end{bmatrix}\begin{bmatrix}\mathbf{y}\\ \mathbf{z}\end{bmatrix}=\begin{bmatrix}0\\ \mathbf{c}\end{bmatrix}, \tag{1.4}$$

while the systems corresponding to (1.3) become

$$(\mathbf{A}^T\mathbf{A}+\delta^2\mathbf{I})\mathbf{x}=\mathbf{A}^T\mathbf{b} \quad \text{and} \quad (\mathbf{A}^T\mathbf{A}+\delta^2\mathbf{I})\mathbf{z}=-\mathbf{c},$$

with $\mathbf{w}=-\delta\mathbf{z}$.

In the remainder of the book, we study a generalization of the above relations and employ iterative numerical methods for least-squares and least-norm problems to solve the saddle-point systems. The reader can verify that if instead of an identity matrix we replace the leading block of the systems (1.1) with a general symmetric and positive-definite $\mathbf{M}$, we change the norm in which the residual is measured in the least-squares problem and in which the solution is measured in the least-norm problem. More precisely, those problems become

$$\underset{\mathbf{x}}{\text{minimize}}\ \tfrac{1}{2}\|\mathbf{A}\mathbf{x}-\mathbf{b}\|^2_{\mathbf{M}^{-1}} \qquad \text{and} \qquad \underset{\mathbf{y}}{\text{minimize}}\ \tfrac{1}{2}\|\mathbf{y}\|^2_{\mathbf{M}} \quad \text{subject to } \mathbf{A}^{\mathsf{T}}\mathbf{y}=\mathbf{c},$$

where the $\mathbf{M}$-norm of a vector is defined as $\|\mathbf{y}\|^2_{\mathbf{M}} = \mathbf{y}^{\mathsf{T}}\mathbf{M}\mathbf{y}$. Similarly, if instead of δI we consider a regularization defined by a symmetric and positive-definite matrix, which we denote $\mathbf{N}^{\frac{1}{2}}$ for convenience, the bottom block of the saddle-point systems becomes $-\mathbf{N}$. The resulting system is called a *symmetric and quasi-definite* system. The iterative methods that we have in mind all belong to the vast category of Krylov methods.

The main contributions of the present work are as follows:

- We propose two families of Krylov methods well suited to the solution of such systems based on a preconditioned variant of the Golub and Kahan (1965) bidiagonalization process. The first family contains methods operating on the normal and Schur-complement equations, including generalizations of well-known methods such as LSQR and LSMR but also a new method named CRAIG-MR aiming to minimize the residual of the Schur-complement equations. The second family follows from a related Lanczos process and contains methods operating directly on the augmented system, generalizing the conjugate-gradient and minimum-residual methods.

- We establish connections between augmented-system and reduced-system methods. In particular, the conjugate-gradient method applied to either system of (1.1) and (1.4) is well defined despite the indefiniteness of the operator.

- We provide an explanation for the often-observed staircase behavior of the residual in the minimum-residual method (Fischer, Ramage, Silvester, and Wathen, 1998).

- We provide explicit stopping criteria for all methods based on estimates of the forward error in an appropriate norm. Orthogonality properties enable us to develop a lower bound estimate on the forward error that is available at essentially no extra computational cost. An upper bound estimate is also available and comes at the cost of a few additional scalar operations per iteration.

Symmetric and quasi-definite (SQD) linear systems have the general form

$$\begin{bmatrix} \mathbf{M} & \mathbf{A} \\ \mathbf{A}^{\mathsf{T}} & -\mathbf{N} \end{bmatrix} \begin{bmatrix} \mathbf{x} \\ \mathbf{y} \end{bmatrix} = \begin{bmatrix} \mathbf{f} \\ \mathbf{g} \end{bmatrix}, \tag{1.5}$$

where $\mathbf{M} \in \mathbb{R}^{n\times n}$ and $\mathbf{N} \in \mathbb{R}^{m\times m}$ are symmetric and positive definite. The matrix $\mathbf{K}$ of (1.5) is said to be SQD. It is always symmetric and indefinite unless $m = 0$ or $n = 0$, in which case it is definite. Vanderbei (1995) shows that $\mathbf{K}$ is *strongly factorizable*; i.e., if $\mathbf{P}$ is any symmetric permutation matrix, the symmetric row and column permutation $\mathbf{PKP}^{\mathsf{T}}$ of $\mathbf{K}$ admits a Cholesky-like factorization $\mathbf{LDL}^{\mathsf{T}}$ with $\mathbf{D}$ diagonal and indefinite. The latter factorization can therefore typically be computed in significantly fewer operations than a traditional symmetric indefinite factorization, such as that of Duff (2004). It will also often have sparser factors because stability pivots are not strictly necessary.

We adopt the following definition of an SQD matrix.

Definition 1.1. *A matrix* $\mathbf{K} \in \mathbb{R}^{(n+m)\times(n+m)}$ *is said to be symmetric quasi-definite (SQD) if* $\mathbf{K} = \mathbf{K}^{\mathsf{T}}$ *and there exists a permutation matrix* $\mathbf{P} \in \mathbb{R}^{(n+m)\times(n+m)}$ *such that* $\mathbf{P}^{\mathsf{T}}\mathbf{K}\mathbf{P}$ *has the form* (1.5).

From this point on, we assume that $n > 0$ and $m > 0$. Among other important properties of (1.5) are that the system is always square, symmetric, indefinite, and nonsingular, regardless of the shape and rank of $\mathbf{A}$. This is in contrast with (1.1), which required $\mathbf{A}$ to be tall and thin and have full column rank. In addition, the inverse of an SQD matrix is itself SQD. For more details, we refer to Vanderbei (1995).

The main objective of this book is to highlight iterative methods for the solution of (1.5) that exploit its structure. The methods in question are generalizations of LSQR (Paige and Saunders, 1982), CRAIG (Craig, 1955), and LSMR (Fong and Saunders, 2011) based on a Golub and Kahan (1965) process occurring in the appropriate metric. In addition, we present a method named CRAIG-MR aiming to minimize the residual of the Schur-complement equations. Those methods are related to implementations of CG (Hestenes and Stiefel, 1952) and MINRES (Paige and Saunders, 1975) defined by a specialized Lanczos processes. The implementation of CG is particularly interesting given the indefiniteness of (1.5).

All methods presented here especially apply to cases where systems involving $\mathbf{M}$ and $\mathbf{N}$ can be solved efficiently. It would be interesting for follow-up research to investigate preconditioning strategies and error analyses related to cases where this assumption is not satisfied. All methods discussed below are dimension agnostic in the sense that each applies regardless of the fact that $m < n$ or $m \geq n$.

One of the early motivating factors for the research that resulted in the present book is that the fields of computational optimization and computational partial-differential equations (PDEs) have a common aspect, that of linear systems with a saddle-point structure. The two fields face similar challenges and progress on the solution of such systems has immediate impact on both fields. We became interested in novel solution methods for (1.5) that did not require any factorization, as such methods would suggest computational procedures for optimization and PDEs that are entirely factorization free. Indeed, one can always apply MINRES or SYMMLQ to (1.5), but it turns out that there are better approaches.

In what follows, we briefly outline a few applications where systems of the form (1.5) arise. Consider the inequality-constrained convex optimization problem

$$\underset{\mathbf{x}\in\mathbb{R}^n}{\text{minimize}}\; f(\mathbf{x}) \quad \text{subject to } c(\mathbf{x}) \geq 0,$$

where f is convex, each component function of c is concave, and all are smooth. Newton-based interior-point methods typically solve direction-finding systems of the form

$$\begin{bmatrix} \nabla_{\mathbf{xx}} L(\mathbf{x},\mathbf{y}) & J(\mathbf{x})^{\mathsf{T}} \\ J(\mathbf{x}) & -Y^{-1}C(\mathbf{x}) \end{bmatrix} \begin{bmatrix} \Delta\mathbf{x} \\ -\Delta\mathbf{y} \end{bmatrix} = - \begin{bmatrix} \nabla_{\mathbf{x}} L(\mathbf{x},\mathbf{y}) \\ c(\mathbf{x}) - \mu Y^{-1}\mathbf{e} \end{bmatrix},$$

where $L(\mathbf{x},\mathbf{y}) = f(\mathbf{x}) - c(\mathbf{x})^{\mathsf{T}}\mathbf{y}$ is the Lagrangian of the problem, $J(\mathbf{x})$ is the Jacobian of the constraints, $Y = \operatorname{diag}(\mathbf{y})$, $C(\mathbf{x}) = \operatorname{diag}(c(\mathbf{x}))$, $\mu > 0$ is a parameter, and $(c(\mathbf{x}),\mathbf{y}) >$

0. Whenever either f is strictly convex or one of the constraint functions is strictly concave, the leading block is positive definite and the system above is SQD. In the optimization context, it is important for the leading block to be positive definite, at least on the nullspace of $J(\mathbf{x})$, so that the search direction Δx is a descent direction for a certain merit function. The technical vocabulary is to say that the saddle-point matrix is *second-order sufficient*.

When the block matrix is not second-order sufficient, one approach is to convexify the problem by adding an increasingly large multiple of the identity to it until it is sufficiently positive definite—see, e.g., Nocedal and Wright (2006). This trial-and-error process can be time consuming and expensive. More importantly, it does not fit well in our framework because solving systems with the leading block is rarely simple, except perhaps in highly structured problems.

An alternative that is computationally promising is to use an approximation $B(\mathbf{x},\mathbf{y}) \approx \nabla_{\mathbf{xx}} L(\mathbf{x},\mathbf{y})$ that both yields a second-order sufficient block matrix and is such that systems involving $B(\mathbf{x},\mathbf{y})$ can be solved efficiently. In optimization, one such approximation is the limited-memory BFGS (L-BFGS).

It is best to not form the approximation explicitly, as that typically results in a dense matrix. Instead, it can be cheaply stored and updated in either factored or inverse form in a way that applying its inverse is cheap and efficient. Moreover, such a $B(\mathbf{x},\mathbf{y})$ is positive definite, and the saddle-point system becomes SQD. We refer the interested reader to Nocedal and Wright (2006) for more information on L-BFGS approximations, as they are not discussed further in the present book.

The mixed finite-element approximation to the globally stabilized Stokes problem in weak form can be stated as

$$\begin{aligned} \int_\Omega \nabla u : \nabla v \, \mathrm{d}x - \int_\Omega p \, \nabla \cdot v \, \mathrm{d}x &= \int_\Omega f v \, \mathrm{d}x \quad \text{for all } v \in V, \\ -\int_\Omega q \, \nabla \cdot u \, \mathrm{d}x - \beta \mathscr{C}(p,q) &= 0 \qquad\qquad \text{for all } q \in Q, \end{aligned} \tag{1.6}$$

where u is the velocity field, p is the pressure field, Ω is the domain, ":" represents the componentwise inner product, V and Q are compatible finite-dimensional test function spaces, $\beta > 0$ is a stabilization parameter, and $\mathscr{C}$ is a stabilization term. After discretization with, e.g., P_1–P_1 triangular elements and continuous linear pressure, the above equations reduce to a linear system of the form (1.5) where $\mathbf{g} = 0$, $\mathbf{A}$ is the gradient matrix, $\mathbf{A}^\mathsf{T}$ is the divergence matrix, $\mathbf{M}$ is the vector-Laplacian matrix, and $\mathbf{N}$ represents the stabilization term. For more details, we refer the interested reader to Silvester and Wathen (1994).

Regularized linear least-squares problems in the Euclidean norm are often cast as systems of the form (1.5) in which $\mathbf{M}$ and $\mathbf{N}$ are multiples of the identity. In this case, the right-hand side has the form $(\mathbf{b}, 0)$ or $(0, \mathbf{c})$. The identification is important because with such a right-hand side, the methods we propose below revolve around an interpretation of (1.5) as a regularized linear least-squares problem in a non-Euclidean metric.

Other applications of regularized linear least squares include Kalman (1960) filters (Bunse-Gerstner, 2012; Strang, 1986) and variational data assimilation (Courtier, 1997). The problem formulation in both applications is very similar. The incremental formulation of the three-dimensional variational data assimilation problem may be stated as

$$\underset{\Delta\mathbf{x}\in\mathbb{R}^n}{\text{minimize}} \; \tfrac{1}{2}\Delta\mathbf{x}^\mathsf{T}\mathbf{B}^{-1}\Delta\mathbf{x} + \tfrac{1}{2}(\mathbf{H}\Delta\mathbf{x}-\mathbf{d})^\mathsf{T}\mathbf{R}^{-1}(\mathbf{H}\Delta\mathbf{x}-\mathbf{d}),$$

where $\Delta\mathbf{x} = \mathbf{x}_0 - \mathbf{x}_b$ is referred to as an *increment* used to obtain an initial climatic model state $\mathbf{x}_0$ from a *background* state $\mathbf{x}_b$—i.e., a state resulting from previous forecasts—$\mathbf{B}$ is the covariance matrix of background errors, $\mathbf{R}$ is the covariance matrix of observation errors, $\mathbf{H}$ represents a linearization of the observation operator, and $\mathbf{d} = \mathbf{y}_0 - \mathbf{H}\mathbf{x}_b$ is the *innovation* vector involving the observation vector $\mathbf{y}_0$. The optimality conditions of the above problem have precisely the form (1.5) with $\mathbf{A} = \mathbf{H}$, $\mathbf{M} = \mathbf{R}$, $\mathbf{N} = \mathbf{B}^{-1}$, $\mathbf{f} = \mathbf{d}$, and $\mathbf{g} = 0$.

Finally, SQD systems have been used in the past to precondition standard symmetric saddle-point systems, i.e., for which $\mathbf{N} = 0$ (Axelsson and Neytcheva, 2003; Perugia and Simoncini, 2000). Benzi, Golub, and Liesen (2005, Section 10.2, pp. 82–83) provide several references and summarize key results including eigenvalue estimates of the preconditioned system.

Notation

Throughout the book, vectors $\mathbf{x}$ and matrices $\mathbf{A}$ are typeset in boldface, while scalars, such as α, appear in lightface. Hilbert spaces are represented in a blackboard font, as in $\mathbb{M}$ and $\mathbb{N}$. Linear operators between Hilbert spaces appear in a cursive font, as in $\mathscr{A}$ and $\mathscr{M}$. We use the notation $\mathbf{I}_k$ to denote the $k \times k$ identity matrix. For conciseness and when the context leaves no possible ambiguity, we will simply denote by $\mathbf{I}$ the identity matrix of appropriate size. For a symmetric positive-definite $n \times n$ matrix $\mathbf{C}$, let $\|\mathbf{u}\|_{\mathbf{C}}^2 = \mathbf{u}^T\mathbf{C}\mathbf{u} = \|\mathbf{C}^{\frac{1}{2}}\mathbf{u}\|_2^2$ be the norm defined by $\mathbf{C}$, where $\mathbf{C}^{\frac{1}{2}}$ is the unique symmetric positive-definite matrix square root of $\mathbf{C}$. The shorthand blkdiag$(\mathbf{C}, \mathbf{D})$ is used to denote the block-diagonal matrix

$$\begin{bmatrix} \mathbf{C} & 0 \\ 0 & \mathbf{D} \end{bmatrix}$$

for any matrices $\mathbf{C}$ and $\mathbf{D}$ of appropriate size. For any $x \in \mathbb{R}$, $\lfloor x \rfloor$ denotes the largest integer $k \leq x$ and $\lceil x \rceil$ denotes the smallest integer $k \geq x$.

Related Work

Existing iterative methods for symmetric indefinite systems, such as MINRES, SYMMLQ (Paige and Saunders, 1975), and SYMMBK (Chandra, 1978), do not exploit the rich quasi-definite structure of (1.5).

Assume we are able to factor $\mathbf{N} = \mathbf{L}\mathbf{D}\mathbf{L}^T$. Upon introducing auxiliary variables $\mathbf{z} := -\mathbf{D}\mathbf{L}^T\mathbf{y}$, it is possible to reformulate (1.5) as the traditional saddle-point system

$$\begin{bmatrix} \mathbf{M} & & \mathbf{A} \\ & \mathbf{D}^{-1} & \mathbf{L}^T \\ \mathbf{A}^T & \mathbf{L} & \end{bmatrix} \begin{bmatrix} \mathbf{x} \\ \mathbf{z} \\ \mathbf{y} \end{bmatrix} = \begin{bmatrix} \mathbf{f} \\ 0 \\ \mathbf{g} \end{bmatrix}. \tag{1.7}$$

Dollar, Gould, Schilders, and Wathen (2006) propose to solve the resulting system by way of the projected conjugate gradient method. This approach requires a projection into the nullspace of $\begin{bmatrix} \mathbf{A}^T & \mathbf{L} \end{bmatrix}$ at each iteration, which may be achieved via a one-time factorization of a projection matrix of the form

$$\begin{bmatrix} \widetilde{\mathbf{M}} & & \mathbf{A} \\ & \mathbf{D}^{-1} & \mathbf{L}^T \\ \mathbf{A}^T & \mathbf{L} & \end{bmatrix},$$

for an appropriate approximation $\widetilde{\mathbf{M}} \approx \mathbf{M}$ such that the above matrix has precisely m negative and $n+m$ positive eigenvalues.

Saunders (1995) derives extended versions of LSQR and CRAIG for the case where $\mathbf{N} = \lambda \mathbf{I}_m$ for some $\lambda \neq 0$ and establishes a connection with CG applied to the corresponding SQD system.

Gill, Saunders, and Shinnerl (1996) provide stability results for the LDL^T factorization of SQD matrices. George, Ikramov, and Kucherov (2000) and George and Ikramov (2000) examine additional properties of SQD matrices, their eigenvalues, and condition number. Korzak (1999) gives the precise spectrum in the case of matrices arising from linear programming.

Benbow (1999) proposes a variant of LSQR similar to what we propose in the present book, only for the case where $\mathbf{N} = 0$ and systems with $\mathbf{M}$ can be solved efficiently.

The definitive reference on solution methods for saddle-point linear systems is given by Benzi, Golub, and Liesen (2005). Although they mention SQD systems, no specialized iterative approach is suggested.

Marcia (2008) proposes a Krylov-type iterative method for general symmetric indefinite systems based on a symmetric indefinite factorization of the tridiagonal matrix generated by a Lanczos process. His method reduces to CG when applied to definite systems and is provably stable on indefinite systems.

Arioli (2013) derives a version of CRAIG for indefinite systems where $\mathbf{N} = 0$ based on the so-called *elliptic singular values* of $\mathbf{A}$.

Chapter 2

Preliminaries

2.1 ▪ Important Note about Notation

For the remainder of the book, we use a notation that slightly differs from that of the introductory chapter. The change of notation states that, rather than viewing an SQD system as following from a linear least-squares problem in which the variables are denoted $\mathbf{x}$ and the residuals $\mathbf{r}$, we mean to view any SQD system, regardless of the application from which it arose, as the optimality conditions of a certain linear least-squares problem that can be deduced from the system. The philosophical difference is that the linear least-squares problem in question was not necessarily apparent in the original application. Thus, in order to focus on the SQD system, we denote its unknowns $(\mathbf{x},\mathbf{y})$, and it will become apparent that it is $\mathbf{y}$ that plays the role of the variables in the least-squares problem, while $\mathbf{x}$ plays the role of the residuals.

2.2 ▪ Linear Least-Squares Problems

In this section, we establish the connections between various symmetric indefinite and SQD linear systems and the solution of certain linear least-squares problems. Note that most of the connections are known. They are repeated here because they are the motivation for the derivation of iterative methods for (1.5). By convexity, all optimality conditions mentioned below are both necessary and sufficient.

The symmetric saddle-point system

$$\begin{bmatrix} \mathbf{M} & \mathbf{A} \\ \mathbf{A}^{\mathsf{T}} & 0 \end{bmatrix} \begin{bmatrix} \mathbf{x} \\ \mathbf{y} \end{bmatrix} = \begin{bmatrix} \mathbf{b} \\ 0 \end{bmatrix}, \tag{2.1}$$

where $\mathbf{M}$ is symmetric and positive definite, represents the optimality conditions of the $\mathbf{M}^{-1}$-norm least-squares problem[1]

$$\underset{\mathbf{y}\in\mathbb{R}^m}{\text{minimize}}\ \tfrac{1}{2}\|\mathbf{A}\mathbf{y}-\mathbf{b}\|^2_{\mathbf{M}^{-1}}, \tag{2.2}$$

where $\mathbf{x} = \mathbf{M}^{-1}(\mathbf{b}-\mathbf{A}\mathbf{y})$ is the residual. The matrix of (2.1) is symmetric indefinite but not SQD. Given the Cholesky decomposition $\mathbf{M} = \mathbf{L}\mathbf{L}^{\mathsf{T}}$, (2.1) also states the optimal-

[1] Sometimes referred to as the *generalized* least-squares problem.

ity conditions of the following *weighted*, or *preconditioned*, least-squares problems

$$\underset{y\in\mathbb{R}^m}{\text{minimize}}\ \tfrac{1}{2}\|\mathbf{L}^{-1}(\mathbf{A}y-\mathbf{b})\|_2^2 \qquad \text{and} \qquad \underset{y\in\mathbb{R}^m}{\text{minimize}}\ \tfrac{1}{2}\|\mathbf{M}^{-\frac{1}{2}}(\mathbf{A}y-\mathbf{b})\|_2^2.$$

Whenever $\mathbf{A}$ does not have full column rank, (2.1) is singular. A typical remedy is to *regularize* the least-square problem, i.e., to change (2.2) to

$$\underset{y\in\mathbb{R}^m}{\text{minimize}}\ \frac{1}{2}\left\|\begin{bmatrix}\mathbf{A}\\ \mathbf{R}\end{bmatrix}y-\begin{bmatrix}\mathbf{b}\\ 0\end{bmatrix}\right\|_{\mathbf{M}_+^{-1}}^2, \qquad \text{where} \qquad \mathbf{M}_+ := \begin{bmatrix}\mathbf{M} & 0\\ 0 & \mathbf{I}_m\end{bmatrix} \tag{2.3}$$

and $\mathbf{R}$ is a square nonsingular matrix of appropriate size. Typically, $\mathbf{R}=\lambda\mathbf{I}_m$ for some regularization parameter $\lambda\in\mathbb{R}$, but other choices are possible. Note that the objective of (2.3) may equivalently be written as

$$\tfrac{1}{2}\|\mathbf{A}y-\mathbf{b}\|_{\mathbf{M}^{-1}}^2+\tfrac{1}{2}\|\mathbf{R}y\|_2^2=\tfrac{1}{2}\|\mathbf{A}y-\mathbf{b}\|_{\mathbf{M}^{-1}}^2+\tfrac{1}{2}\|y\|_{\mathbf{R}^T\mathbf{R}}^2.$$

The optimality conditions of (2.3) may then be written as the general SQD system

$$\begin{bmatrix}\mathbf{M} & \mathbf{A}\\ \mathbf{A}^T & -\mathbf{R}^T\mathbf{R}\end{bmatrix}\begin{bmatrix}x\\ y\end{bmatrix}=\begin{bmatrix}\mathbf{b}\\ 0\end{bmatrix}.$$

Similarly, a positive-definite matrix of the form $\mathbf{N}^{\frac{1}{2}}$ can be used in place of $\mathbf{R}$, and this leads to (1.5) with $\mathbf{f}=\mathbf{b}$ and $\mathbf{g}=0$. Additionally, (1.5) with $\mathbf{f}=\mathbf{b}$ and $\mathbf{g}=0$ represents the optimality conditions of the $\mathbf{E}_+^{-1}$-norm regularized problem

$$\underset{y\in\mathbb{R}^m}{\text{minimize}}\ \frac{1}{2}\left\|\begin{bmatrix}\mathbf{A}\\ \mathbf{I}_m\end{bmatrix}y-\begin{bmatrix}\mathbf{b}\\ 0\end{bmatrix}\right\|_{\mathbf{E}_+^{-1}}^2, \qquad \text{where} \qquad \mathbf{E}_+ := \begin{bmatrix}\mathbf{M} & 0\\ 0 & \mathbf{N}^{-1}\end{bmatrix}. \tag{2.4}$$

Equivalently, if $\mathbf{M}=\mathbf{L}\mathbf{L}^T$ and $\mathbf{N}=\mathbf{R}^T\mathbf{R}$, (1.5) represents the optimality conditions of the weighted regularized problem

$$\underset{y\in\mathbb{R}^m}{\text{minimize}}\ \frac{1}{2}\left\|\begin{bmatrix}\mathbf{L}^{-1} & 0\\ 0 & \mathbf{R}\end{bmatrix}\left(\begin{bmatrix}\mathbf{A}\\ \mathbf{I}_m\end{bmatrix}y-\begin{bmatrix}\mathbf{b}\\ 0\end{bmatrix}\right)\right\|_2^2. \tag{2.5}$$

A similar interpretation is derived when $\mathbf{L}$ and $\mathbf{R}$ are replaced by $\mathbf{M}^{\frac{1}{2}}$ and $\mathbf{N}^{\frac{1}{2}}$, respectively.

At this point, it might seem attractive to simply employ LSQR to solve either (2.4) or (2.5). This would, however, require that we either compute $\mathbf{L}$ or solve systems with $\mathbf{M}^{\frac{1}{2}}$ and that we compute one of $\mathbf{R}$ and $\mathbf{N}^{\frac{1}{2}}$. Moreover, those must be computed accurately. Fortunately, there is an alternative in applications where solving systems with matrices $\mathbf{M}$ and $\mathbf{N}$ can be performed efficiently. We now give a few examples that arise in practice.

In regularized interior-point methods for linear programming—see, e.g., Friedlander and Orban (2012)—the matrices $\mathbf{M}$ and $\mathbf{N}$ are diagonal, and solving systems with those matrices is therefore trivial. In interior-point methods for nonlinear programming in which the Hessian of the Lagrangian is approximated by a limited-memory quasi-Newton matrix in inverse form, solving systems with $\mathbf{M}$ is cheap because the limited-memory approximation represents $\mathbf{M}^{-1}$. A matrix-vector product is thus all that is required. Similarly, if the limited-memory quasi-Newton matrix is updated in factored form, solving systems with $\mathbf{M}$ is cost effective.

In fluid dynamics applications, the discretization of Darcy's law for incompressible flow in a saturated medium gives rise to a badly scaled matrix $\mathbf{M}$ but for which diagonal preconditioning will leave only a few clusters of eigenvalues independently of the mesh size (Wathen, 1987). It can thus be expected that the conjugate gradient method with diagonal preconditioner will converge quickly.

Without loss of generality, we assume from now on that the right-hand side of (1.5) has $\mathbf{f} = \mathbf{b}$ and $\mathbf{g} = 0$. Reduction to this situation is always possible, though admittedly at some cost, by first finding any $(\mathbf{x}^0, \mathbf{y}^0)$ satisfying $\mathbf{A}^T\mathbf{x}^0 - \mathbf{N}\mathbf{y}^0 = \mathbf{g}$ and setting $\mathbf{b} = \mathbf{f} - \mathbf{M}\mathbf{x}^0 - \mathbf{A}\mathbf{y}^0$. A variety of iterative methods can be used to identify such $(\mathbf{x}^0, \mathbf{y}^0)$. For instance, the minimum norm problem

$$\underset{\mathbf{x},\mathbf{y}}{\text{minimize}} \ \tfrac{1}{2}\left(\|\mathbf{x}\|_2^2 + \|\mathbf{y}\|_2^2\right) \quad \text{subject to } \mathbf{A}\mathbf{x} - \mathbf{N}\mathbf{y} = \mathbf{g}$$

can be solved with either the standard LSQR (Paige and Saunders, 1982) or CRAIG (Craig, 1955). Because of our assumption that solving systems with $\mathbf{N}$ can be done easily and efficiently, a simpler solution consists in setting $\mathbf{x}^0 := 0$ and solving $\mathbf{N}\mathbf{y}^0 = -\mathbf{g}$ for $\mathbf{y}^0$.

Consider (1.5) with $\mathbf{f} = \mathbf{b}$ and $\mathbf{g} = 0$. If we eliminate $\mathbf{x}$ from the first equation, we see that the $\mathbf{y}$ component of the solution must satisfy

$$(\mathbf{A}^T\mathbf{M}^{-1}\mathbf{A} + \mathbf{N})\mathbf{y} = \mathbf{A}^T\mathbf{M}^{-1}\mathbf{b}. \tag{2.6}$$

We refer to (2.6) as the *normal equations*. Similarly, if we eliminate $\mathbf{y}$ from the second equation of (1.5), we see that the $\mathbf{x}$ component must satisfy

$$(\mathbf{A}\mathbf{N}^{-1}\mathbf{A}^T + \mathbf{M})\mathbf{x} = \mathbf{b}, \tag{2.7}$$

to which we refer as the *Schur-complement equations*. Those equations are not, strictly speaking, normal equations, as they do not directly describe the optimality conditions of a linear least-squares problem. As becomes apparent in later sections, the matrices of (2.6) and (2.7) play an important role in our iterative methods in that they define energy norms suitable for measuring direct errors.

We close this section by mentioning that (2.4) is always equivalent to the least-norm problem

$$\underset{\mathbf{x},\mathbf{y}}{\text{minimize}} \ \tfrac{1}{2}(\|\mathbf{x}\|_{\mathbf{M}}^2 + \|\mathbf{y}\|_{\mathbf{N}}^2) \quad \text{subject to } \mathbf{M}\mathbf{x} + \mathbf{A}\mathbf{y} = \mathbf{b}. \tag{2.8}$$

Indeed, the Lagrange multipliers associated with the equality constraints are precisely equal to $\mathbf{x}$. It is only when regularization is present that the least-squares problem and the least-norm problems are equivalent.

2.3 ▪ Hilbert Space Setting

Let $\mathbf{H} \in \mathbb{R}^{k\times k}$ be symmetric and positive definite. Then $\mathbb{R}^k$ endowed with the scalar product $\mathbf{u}^T\mathbf{H}\mathbf{v}$ is a Hilbert space. Conversely, let $\mathbb{H}$ be a k-dimensional Hilbert space with basis $\{\phi_j\}_{j=1,\ldots,k}$ and equipped with a scalar product $(u,v)_{\mathbb{H}}$. Then $\mathbb{H}$ is isometric to $\mathbb{R}^k$ with a scalar product determined by the Gramian matrix $\mathbf{H}$, i.e., $\mathbf{H}_{ij} := (\phi_i, \phi_j)_{\mathbb{H}}$. Indeed, upon decomposing $u = \sum_j u_j\phi_j$ and $v = \sum_j v_j\phi_j$, we have $(u,v)_{\mathbb{H}} = \mathbf{u}^T\mathbf{H}\mathbf{v}$. Owing to the Riesz theorem (Brézis, 1983), the dual space $\mathbb{H}^\star$ of

$\mathbb{H}$ is itself a Hilbert space with a scalar product induced by $\mathbf{H}^{-1}$. In particular, the operator

$$\mathscr{H} : \mathbb{H} \to \mathbb{H}^\star \qquad \langle \mathscr{H} u, v \rangle_{\mathbb{H}^\star,\mathbb{H}} := \mathbf{v}^\mathsf{T}\mathbf{H}\mathbf{u}$$

is self-adjoint and strictly positive and therefore invertible. Furthermore, the basis $\{\phi_i\}$ is made of the columns of $\mathbf{H}$, and the corresponding basis $\{\psi_i\}$ of $\mathbb{H}^\star$ is made of the columns of $\mathbf{H}^{-1}$. Hereafter, all our Hilbert spaces are finite-dimensional.

Given $z \in \mathbb{H}^\star$, we have

$$\langle z, u \rangle_{\mathbb{H}^\star,\mathbb{H}} = \mathbf{z}^\mathsf{T}\mathbf{u} = \mathbf{z}^\mathsf{T}\mathbf{H}^{-1}\mathbf{H}\mathbf{u} = (\mathbf{u}, \mathbf{H}^{-1}\mathbf{z})_{\mathbb{H}},$$

and we have that $\mathbf{w} = \mathbf{H}^{-1}\mathbf{z}$ is the representation of the Riesz vector $w = \sum_j w_j \phi_j \in \mathbb{H}$. Let $\mathscr{K} : \mathbb{H} \to \mathbb{F}$ be an operator between the Hilbert spaces $\mathbb{H}$ and $\mathbb{F}$. Its adjoint operator $\mathscr{K}^\star : \mathbb{F}^\star \to \mathbb{H}^\star$ is defined (Brézis, 1983) by

$$\langle \mathscr{K}^\star v, u \rangle_{\mathbb{H}^\star,\mathbb{H}} := \langle v, \mathscr{K} u \rangle_{\mathbb{F}^\star,\mathbb{F}} \quad \text{for all } v \in \mathbb{F}^\star, u \in \mathbb{H}.$$

Therefore, we have

$$\langle \mathscr{K}^\star v, u \rangle_{\mathbb{H}^\star,\mathbb{H}} = (\mathbf{H}^{-1}\mathbf{v}, \mathbf{K}\mathbf{u})_{\mathbb{H}} = \mathbf{u}^\mathsf{T}\mathbf{K}^\mathsf{T}\mathbf{v}, \tag{2.9}$$

where $\mathbf{K}$ is a matrix representation of $\mathscr{K}$. Finally, if we assume that $\mathbb{F} = \mathbb{H}^\star$, then the "normal equations operator" is

$$\mathscr{K}^\star \circ \mathscr{H}^{-1} \circ \mathscr{K} : \mathbb{H} \to \mathbb{H}^\star,$$

and it is represented by the matrix $\mathbf{K}^\mathsf{T}\mathbf{H}^{-1}\mathbf{K}$. If $\mathbf{K}^\mathsf{T} = \mathbf{K}$, then $\mathscr{K}$ is self-adjoint. Moreover, the operator

$$\mathscr{H}^{-1} \circ \mathscr{K} : \mathbb{H} \to \mathbb{H} \tag{2.10}$$

maps $\mathbb{H}$ into itself. Therefore, we can define its powers $(\mathscr{H}^{-1} \circ \mathscr{K})^i$ as the operators represented by the matrices $(\mathbf{H}^{-1}\mathbf{K})^i$ for $i \geq 0$.

Let us consider now the Hilbert spaces

$$\mathbb{M} := (\mathbb{R}^n, \|\cdot\|_{\mathbf{M}}), \qquad \mathbb{N} := (\mathbb{R}^m, \|\cdot\|_{\mathbf{N}}),$$

their duals

$$\mathbb{M}^\star := (\mathbb{R}^n, \|\cdot\|_{\mathbf{M}^{-1}}), \qquad \mathbb{N}^\star := (\mathbb{R}^m, \|\cdot\|_{\mathbf{N}^{-1}}),$$

and assume that

$$\mathbf{K} := \begin{bmatrix} \mathbf{M} & \mathbf{A} \\ \mathbf{A}^\mathsf{T} & -\mathbf{N} \end{bmatrix} \tag{2.11}$$

with the corresponding operator

$$\mathscr{K} : \mathbb{M} \times \mathbb{N} \to \mathbb{M}^\star \times \mathbb{N}^\star.$$

The norm and scalar product in $\mathbb{M} \times \mathbb{N}$ are induced by the block-diagonal matrix

$$\mathbf{H} = \begin{bmatrix} \mathbf{M} & \\ & \mathbf{N} \end{bmatrix}. \tag{2.12}$$

Let the matrix $\mathbf{A}$ represent the linear operator $\mathscr{A} : \mathbb{N} \to \mathbb{M}^\star$ with respect to the canonical bases. For $y \in \mathbb{N}$, $\mathscr{A} y$ may be considered as a linear operator defined on $\mathbb{M}$ via the relation

$$\langle \mathscr{A} y, u \rangle_{\mathbb{M}^\star,\mathbb{M}} := (\mathbf{u}, \mathbf{M}^{-1}\mathbf{A}\mathbf{y})_{\mathbf{M}} = \mathbf{u}^\mathsf{T}\mathbf{A}\mathbf{y} \quad \text{for all } u \in \mathbb{M},$$

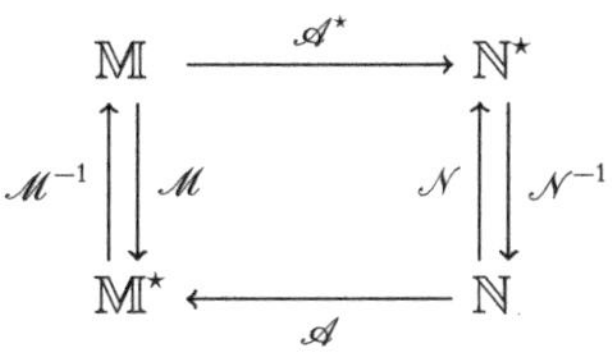

Figure 2.1. *Commutative diagram between the relevant Hilbert spaces.*

where $\langle\cdot,\cdot\rangle_{\mathbb{M}^\star,\mathbb{M}}$ is the duality pairing between $\mathbb{M}$ and its dual. It is now clear that the appropriate norm to measure the residual $\mathbf{b}-\mathbf{Ay}$ is the $\mathbf{M}^{-1}$-norm. Let $\mathscr{A}^\star:\mathbb{M}\to\mathbb{N}^\star$ denote the adjoint operator of $\mathscr{A}$, i.e.,

$$\langle \mathscr{A}^\star u, y\rangle_{\mathbb{N}^\star,\mathbb{N}} := (\mathbf{y}, \mathbf{N}^{-1}\mathbf{A}^\mathsf{T}\mathbf{u})_\mathbf{N} = \mathbf{y}^\mathsf{T}\mathbf{A}^\mathsf{T}\mathbf{u} \quad \text{for all } y\in\mathbb{N}.$$

Finally, let $\mathscr{M}:\mathbb{M}\to\mathbb{M}^\star$, $\mathscr{N}:\mathbb{N}\to\mathbb{N}^\star$, and their inverses $\mathscr{M}^{-1}:\mathbb{M}^\star\to\mathbb{M}$ and $\mathscr{N}^{-1}:\mathbb{N}^\star\to\mathbb{N}$ denote the linear operators whose representations are the matrices $\mathbf{M}$ and $\mathbf{N}$ and their inverses.

We define the operator

$$(\mathscr{A}^\star\circ\mathscr{M}^{-1}\circ\mathscr{A})+\mathscr{N}:\mathbb{N}\to\mathbb{N}^\star$$

as the *normal operator*. This operator appears in the normal equations (2.6). The appropriate norm to measure the residual of the normal equations is the $\mathbf{N}^{-1}$-norm. Similarly, we call the operator

$$(\mathscr{A}\circ\mathscr{N}^{-1}\circ\mathscr{A}^\star)+\mathscr{M}:\mathbb{M}\to\mathbb{M}^\star$$

the *Schur-complement operator*. The residual of the Schur-complement equations is measured in the $\mathbf{M}^{-1}$-norm. The situation is summarized in the commutative diagram of Figure 2.1.

The next sections will regularly refer to Figure 2.1, as they provide appropriate norms in which various quantities such as direct errors $\mathbf{y}_*-\mathbf{y}_k$ and residuals $\mathbf{x}_*-\mathbf{x}_k$ should be measured. The commutative diagram proves to be a consistently useful tool in understanding why such norms are appropriate.

2.4 ▪ Properties of SQD Matrices and Their Krylov Spaces

2.4.1 ▪ Eigenvalues

From Sylvester's law of inertia, the congruence relation

$$\begin{bmatrix}\mathbf{M} & \mathbf{A}\\ \mathbf{A}^\mathsf{T} & -\mathbf{N}\end{bmatrix} = \begin{bmatrix}\mathbf{I}_n & \\ \mathbf{A}^\mathsf{T}\mathbf{M}^{-1} & \mathbf{I}_m\end{bmatrix}\begin{bmatrix}\mathbf{M} & \\ & -(\mathbf{A}^\mathsf{T}\mathbf{M}^{-1}\mathbf{A}+\mathbf{N})\end{bmatrix}\begin{bmatrix}\mathbf{I}_n & \mathbf{M}^{-1}\mathbf{A}\\ & \mathbf{I}_m\end{bmatrix} \tag{2.13}$$

shows that the matrix of (1.5) always possesses precisely n positive and m negative eigenvalues. Note that a second possible decomposition illustrating this result is

$$\begin{bmatrix}\mathbf{M} & \mathbf{A}\\ \mathbf{A}^\mathsf{T} & -\mathbf{N}\end{bmatrix} = \begin{bmatrix}\mathbf{I}_n & -\mathbf{A}\mathbf{N}^{-1}\\ & \mathbf{I}_m\end{bmatrix}\begin{bmatrix}\mathbf{A}\mathbf{N}^{-1}\mathbf{A}^\mathsf{T}+\mathbf{M} & \\ & -\mathbf{N}\end{bmatrix}\begin{bmatrix}\mathbf{I}_n & \\ -\mathbf{N}^{-1}\mathbf{A}^\mathsf{T} & \mathbf{I}_m\end{bmatrix}. \tag{2.14}$$

The result below is more precise. Consider centered preconditioning of (1.5) with $f = b$ and $g = 0$:

$$\begin{bmatrix} M^{-\frac{1}{2}} & \\ & N^{-\frac{1}{2}} \end{bmatrix} \begin{bmatrix} M & A \\ A^T & -N \end{bmatrix} \begin{bmatrix} M^{-\frac{1}{2}} & \\ & N^{-\frac{1}{2}} \end{bmatrix} \begin{bmatrix} M^{\frac{1}{2}}x \\ N^{\frac{1}{2}}y \end{bmatrix} = \begin{bmatrix} M^{-\frac{1}{2}}b \\ 0 \end{bmatrix}. \tag{2.15}$$

It is straightforward to verify that the matrix of the previous system is

$$\overline{K} := \begin{bmatrix} I_n & \overline{A} \\ \overline{A}^T & -I_m \end{bmatrix} \quad \text{with} \quad \overline{A} := M^{-\frac{1}{2}} A N^{-\frac{1}{2}}. \tag{2.16}$$

The next result gives the eigenvalues of $\overline{K}$. It is a special case of Saunders (1995, Result 2).

Theorem 2.1 (Eigenvalues of a SQD Matrix). *Suppose $\overline{A}$ has rank $p \le \min(m, n)$ with nonzero singular values $\sigma_1, \dots, \sigma_p$. The eigenvalues of $\overline{K}$ are*

- $\lambda = +1$ *with multiplicity* $n - p$,
- $\lambda = -1$ *with multiplicity* $m - p$,
- $\lambda = \pm\sqrt{1 + \sigma_k^2}$, $k = 1, \dots, p$.

The scalars σ_k are the singular values of $M^{-\frac{1}{2}} A N^{-\frac{1}{2}}$, i.e., the elliptic singular values of A. A similar result clearly also holds if we replace $M^{\frac{1}{2}}$ with L and $N^{\frac{1}{2}}$ with R, where $M = LL^T$ and $N = R^T R$.

Theorem 2.1 implies that the spectrum of $\overline{K}$ is symmetric; i.e., if λ is an eigenvalue of $\overline{K}$, then $-\lambda$ is another. An important result related to operators with symmetric spectrum due to Fischer (2011, Theorem 6.9.9) proves to be instrumental to our analysis.

Consider K, H, and $\overline{K}$ as defined in (2.11), (2.12), and (2.16) and observe that

$$K = H^{\frac{1}{2}} \overline{K} H^{\frac{1}{2}}. \tag{2.17}$$

By direct computation,

$$\overline{K}^2 = \begin{bmatrix} I_n + \overline{A}\,\overline{A}^T & \\ & I_m + \overline{A}^T \overline{A} \end{bmatrix} := \overline{D}. \tag{2.18}$$

From (2.18) and the symmetry of $\overline{K}$, we have the following properties:

$$\overline{K}^{-1} = \overline{D}^{-1} \overline{K} = \overline{K}\,\overline{D}^{-1}, \tag{2.19}$$

$$\overline{K}\,\overline{D} = \overline{K}^3 = \overline{D}\,\overline{K}, \tag{2.20}$$

$$K H^{-1} K = H^{\frac{1}{2}} \overline{D} H^{\frac{1}{2}} = \begin{bmatrix} M + A N^{-1} A^T & \\ & N + A^T M^{-1} A \end{bmatrix} := D. \tag{2.21}$$

Note that D contains both the normal equations matrix (2.6) and the Schur-complement equations (2.7) matrix. Finally, from (2.21), we have

$$K^{-1} = D^{-1} K H^{-1} = H^{-1} K D^{-1}. \tag{2.22}$$

Because (2.20) says that $\bar{\mathbf{D}}$ and $\bar{\mathbf{K}}$ commute, both matrices can be simultaneously diagonalized. In addition, (2.17) and (2.21) imply that $\mathbf{D}$ and $\mathbf{K}$ can be simultaneously diagonalized by the solutions of the generalized eigenvalue problem

$$\mathbf{K}\mathbf{z} = \lambda_j \mathbf{H}\mathbf{z},$$

where the λ_j, $j = 1, \ldots, p = \operatorname{rank}(\bar{\mathbf{A}})$ are the same eigenvalues presented in Theorem 2.1. Again, the entire discussion above remains if we replace $\mathbf{H}^{\frac{1}{2}}$ by the Cholesky factor of $\mathbf{H}$.

2.4.2 ▪ Krylov subspaces

We denote

$$\mathscr{K}_i(\bar{\mathbf{K}}, \bar{\mathbf{z}}) = \operatorname{Range}\left\{\bar{\mathbf{z}}, \bar{\mathbf{K}}\bar{\mathbf{z}}, \ldots, \bar{\mathbf{K}}^{i-1}\bar{\mathbf{z}}, \bar{\mathbf{K}}^{i}\bar{\mathbf{z}}\right\} \tag{2.23}$$

the ith Krylov subspace generated by $\bar{\mathbf{K}}$ and a vector $\bar{\mathbf{z}}$. Note that $\mathscr{K}_i(\bar{\mathbf{K}}, \bar{\mathbf{z}})$ is also the ith Krylov subspace generated by $\mathbf{K}$ symmetrically preconditioned by $\mathbf{H}^{\frac{1}{2}}$ and the vector $\mathbf{z} = \mathbf{H}^{\frac{1}{2}}\bar{\mathbf{z}}$. Moreover, taking into account (2.10), the discussion in Section 2.3, and (2.17), we have

$$\mathbf{H}^{-1}\mathbf{K} = \mathbf{H}^{-\frac{1}{2}}\bar{\mathbf{K}}\mathbf{H}^{\frac{1}{2}} \tag{2.24}$$

and

$$\mathscr{K}_i(\mathbf{H}^{-1}\mathbf{K}, \mathbf{w}) = \mathbf{H}^{-\frac{1}{2}}\mathscr{K}_i(\bar{\mathbf{K}}, \mathbf{z}), \qquad \text{where} \qquad \mathbf{w} = \mathbf{H}^{-\frac{1}{2}}\mathbf{z}. \tag{2.25}$$

Owing to the symmetry of the spectrum of $\bar{\mathbf{K}}$—see Theorem 2.1—it is known (see Fischer (2011, Theorem 6.9.9) and Freund, Golub, and Nachtigal (1991)) that Lanczos-based algorithms such as MINRES perform redundant iterations. Taking into account the structure of $\bar{\mathbf{K}}$, it is possible to be more precise.

From (2.18) and (2.20), we have, for all $k \geq 0$,

$$\bar{\mathbf{K}}^{2k} = \bar{\mathbf{D}}^{k} \qquad \text{and} \qquad \bar{\mathbf{K}}^{2k+1} = \bar{\mathbf{K}}\bar{\mathbf{D}}^{k} = \bar{\mathbf{D}}^{k}\bar{\mathbf{K}}. \tag{2.26}$$

Consequently, for all $i \geq 0$, the Krylov subspace $\mathscr{K}_i(\bar{\mathbf{K}}, \bar{\mathbf{z}})$ can be written as the direct sum

$$\begin{aligned} \mathscr{K}_i(\bar{\mathbf{K}}, \bar{\mathbf{z}}) &= \mathscr{K}_{\lfloor i/2 \rfloor}(\bar{\mathbf{D}}, \bar{\mathbf{z}}) \oplus \mathscr{K}_{\lfloor i/2 \rfloor}(\bar{\mathbf{D}}, \bar{\mathbf{K}}\bar{\mathbf{z}}) \\ &= \mathscr{K}_{\lfloor i/2 \rfloor}(\bar{\mathbf{D}}, \bar{\mathbf{z}}) \oplus \bar{\mathbf{K}}\mathscr{K}_{\lfloor i/2 \rfloor}(\bar{\mathbf{D}}, \bar{\mathbf{z}}). \end{aligned} \tag{2.27}$$

Let $\bar{\mathbf{D}}_1$ and $\bar{\mathbf{D}}_2$ be defined such that $\bar{\mathbf{D}} = \operatorname{blkdiag}(\bar{\mathbf{D}}_1, \bar{\mathbf{D}}_2)$. Then, if $\bar{z} = (\bar{z}_1, \bar{z}_2)$, we have

$$\mathscr{K}_i(\bar{\mathbf{D}}, \bar{\mathbf{z}}) = \begin{bmatrix} \mathscr{K}_i(\bar{\mathbf{D}}_1, \bar{\mathbf{z}}_1) \\ 0 \end{bmatrix} \oplus \begin{bmatrix} 0 \\ \mathscr{K}_i(\bar{\mathbf{D}}_2, \bar{\mathbf{z}}_2) \end{bmatrix} \tag{2.28}$$

and

$$\bar{\mathbf{K}}\mathscr{K}_i(\bar{\mathbf{D}}, \bar{\mathbf{z}}) = \begin{bmatrix} \mathscr{K}_i(\bar{\mathbf{D}}_1, \bar{\mathbf{z}}_1) \\ \bar{\mathbf{A}}^{\mathsf{T}}\mathscr{K}_i(\bar{\mathbf{D}}_1, \bar{\mathbf{z}}_1) \end{bmatrix} \oplus \begin{bmatrix} \bar{\mathbf{A}}\mathscr{K}_i(\bar{\mathbf{D}}_2, \bar{\mathbf{z}}_2) \\ -\mathscr{K}_i(\bar{\mathbf{D}}_2, \bar{\mathbf{z}}_2) \end{bmatrix} \tag{2.29}$$

$$= \begin{bmatrix} \mathscr{K}_i(\bar{\mathbf{D}}_1, \bar{\mathbf{z}}_1) \\ \mathscr{K}_i(\bar{\mathbf{D}}_2, \bar{\mathbf{A}}^{\mathsf{T}}\bar{\mathbf{z}}_1) \end{bmatrix} \oplus \begin{bmatrix} \mathscr{K}_i(\bar{\mathbf{D}}_1, \bar{\mathbf{A}}\bar{\mathbf{z}}_2) \\ -\mathscr{K}_i(\bar{\mathbf{D}}_2, \bar{\mathbf{z}}_2) \end{bmatrix}. \tag{2.30}$$

In particular, if we choose $\mathbf{z} = (\mathbf{b}, 0)$ or $(0, \mathbf{g})$, we have, respectively,

$$\mathscr{K}_i\left(\bar{\mathbf{D}}, \begin{bmatrix} \mathbf{b} \\ 0 \end{bmatrix}\right) = \begin{bmatrix} \mathscr{K}_i(\bar{\mathbf{D}}_1, \mathbf{b}) \\ 0 \end{bmatrix} \quad \text{and} \quad \mathscr{K}_i\left(\bar{\mathbf{D}}, \begin{bmatrix} 0 \\ \mathbf{g} \end{bmatrix}\right) = \begin{bmatrix} 0 \\ \mathscr{K}_i(\bar{\mathbf{D}}_2, \mathbf{g}) \end{bmatrix}.$$

With $\mathbf{z} = (\mathbf{b}, 0)$, we finally obtain

$$\mathscr{K}_i\left(\bar{\mathbf{K}}, \begin{bmatrix} \mathbf{b} \\ 0 \end{bmatrix}\right) = \begin{bmatrix} \mathscr{K}_{\lfloor i/2 \rfloor}(\bar{\mathbf{D}}_1, \mathbf{b}) \\ 0 \end{bmatrix} \oplus \begin{bmatrix} 0 \\ \mathscr{K}_{\lfloor i/2 \rfloor + 1}(\bar{\mathbf{D}}_2, \mathbf{A}^\mathsf{T}\mathbf{b}) \end{bmatrix}.$$

Chapter 3

Overview of Existing Direct and Iterative Methods

This chapter covers some background material and known properties of SQD matrices. It also somewhat anticipates certain concepts and should be seen as a broad overview.

Vanderbei (1995) established that for any permutation matrix $\mathbf{P}$, the symmetric permutation of the SQD matrix

$$\mathbf{K} = \begin{bmatrix} \mathbf{M} & \mathbf{A} \\ \mathbf{A}^\mathsf{T} & -\mathbf{N} \end{bmatrix}, \tag{3.1}$$

with $\mathbf{M}$ and $\mathbf{N}$ symmetric and positive definite, can be factorized without pivoting, i.e.,

$$\mathbf{P}^\mathsf{T}\mathbf{K}\mathbf{P} = \mathbf{L}\mathbf{D}\mathbf{L}^\mathsf{T}, \tag{3.2}$$

where $\mathbf{L}$ is lower triangular with unit diagonal and $\mathbf{D}$ is diagonal. This property, called *strong factorizability*, is particularly relevant when $\mathbf{K}$ is large and sparse. As we illustrate below, not all permutations yield stable factorizations or even sparse factorizations. However, exactly as in the case of definite matrices, strong factorizability implies that we are free to select a permutation that attempts to minimize the fill-in in the factor $\mathbf{L}$.

Vanderbei's proof of (3.2) is in two parts. The first part consists in establishing that $\mathbf{K}$ is always nonsingular, which can be verified directly by showing that its nullspace is reduced to $\{0\}$, but it is instructive to appeal to a classic result on Schur complements called Schur's formula (Zhang, 2005, Theorem 1.1). We apply Schur's formula in Section 3.1 and provide further comments on the inverse of a SQD matrix. We return to strong factorizability in Section 3.2.

3.1 ▪ Inverse of a SQD Matrix

For conciseness, we denote

$$\mathbf{K}/\mathbf{M} := -\mathbf{N} - \mathbf{A}^\mathsf{T}\mathbf{M}^{-1}\mathbf{A}$$

the Schur complement of $\mathbf{M}$ in $\mathbf{K}$ and

$$\mathbf{K}/(-\mathbf{N}) := \mathbf{M} + \mathbf{A}\mathbf{N}^{-1}\mathbf{A}^\mathsf{T}$$

the Schur complement of $-\mathbf{N}$ in $\mathbf{K}$. Note that we have already encountered those Schur complements in the matrix $\mathbf{D}$ of (2.21). Schur's formula states that $\mathbf{K}$ is nonsingular if and only if both $\mathbf{M}$ and $\mathbf{K}/\mathbf{M}$ are nonsingular. Indeed, in the present case, $\mathbf{K}/\mathbf{M}$ is negative definite. Schur's formula may also be used to establish that $\mathbf{K}$ is nonsingular if and only if both $-\mathbf{N}$ and $\mathbf{K}/(-\mathbf{N})$ are nonsingular. And, indeed, $\mathbf{K}/(-\mathbf{N})$ is positive definite.

As it turns out, Vanderbei (1995, Theorem 1.1) establishes the more precise result that the inverse of a SQD matrix is itself SQD. The result can be seen as an application of Babachiewicz's formula for the inverse of a partitioned matrix in terms of Schur complements (Zhang, 2005, Theorem 1.2) and its many equivalent forms:

$$\begin{bmatrix} \mathbf{M} & \mathbf{A} \\ \mathbf{A}^\mathsf{T} & -\mathbf{N} \end{bmatrix}^{-1} = \begin{bmatrix} (\mathbf{K}/(-\mathbf{N}))^{-1} & -\mathbf{M}^{-1}\mathbf{A}(\mathbf{K}/\mathbf{M})^{-1} \\ -(\mathbf{K}/\mathbf{M})^{-1}\mathbf{A}^\mathsf{T}\mathbf{M}^{-1} & (\mathbf{K}/\mathbf{M})^{-1} \end{bmatrix}, \tag{3.3}$$

which is visibly SQD because $\mathbf{K}/(-\mathbf{N})$ is positive definite and $\mathbf{K}/\mathbf{M}$ is negative definite.

The Aitken block-diagonalization formula states the congruence relation

$$\begin{bmatrix} \mathbf{I} & \\ -\mathbf{A}\mathbf{M}^{-1} & \mathbf{I} \end{bmatrix} \begin{bmatrix} \mathbf{M} & \mathbf{A} \\ \mathbf{A}^\mathsf{T} & -\mathbf{N} \end{bmatrix} \begin{bmatrix} \mathbf{I} & -\mathbf{M}^{-1}\mathbf{A} \\ & \mathbf{I} \end{bmatrix} = \begin{bmatrix} \mathbf{M} & \\ & \mathbf{K}/\mathbf{M} \end{bmatrix}$$

and, by way of Sylvester's law of inertia, establishes that the inertia of $\mathbf{K}$ is the sum of the inertias of $\mathbf{M}$ and $\mathbf{K}/\mathbf{M}$, i.e., $(n, m, 0)$. Formulae for the inverse of $\mathbf{K}$ may be derived from Aitken's formula.

Another formula for the inverse reveals that the leading block in (3.3) is

$$(\mathbf{K}/(-\mathbf{N}))^{-1} = \mathbf{M}^{-1} + \mathbf{M}^{-1}\mathbf{A}(\mathbf{K}/\mathbf{M})^{-1}\mathbf{A}^\mathsf{T}\mathbf{M}^{-1}. \tag{3.4}$$

A disadvantage of the latter expression is that it is not visibly positive definite because it is the sum of a positive-definite matrix and a negative-definite matrix. An advantage is that it shows that the leading block of the inverse is a scaled contraction, as we now establish.

First note that

$$\mathbf{K} = \begin{bmatrix} \mathbf{M}^{\frac{1}{2}} & \\ & \mathbf{N}^{\frac{1}{2}} \end{bmatrix} \begin{bmatrix} \mathbf{I} & \overline{\mathbf{A}} \\ \overline{\mathbf{A}}^\mathsf{T} & -\mathbf{I} \end{bmatrix} \begin{bmatrix} \mathbf{M}^{\frac{1}{2}} & \\ & \mathbf{N}^{\frac{1}{2}} \end{bmatrix}, \qquad \overline{\mathbf{A}} := \mathbf{M}^{-\frac{1}{2}}\mathbf{A}\mathbf{N}^{-\frac{1}{2}}.$$

Let $\overline{\mathbf{A}}^\mathsf{T} = \mathbf{U}\Sigma\mathbf{V}^\mathsf{T}$ be the full singular value decomposition of $\overline{\mathbf{A}}^\mathsf{T}$, where $\mathbf{U}$ and $\mathbf{V}$ are square and orthogonal and $\Sigma = \mathrm{diag}(\sigma_i)$. We have the *canonical* decomposition

$$\mathbf{K} = \mathbf{W}\mathring{\mathbf{K}}\mathbf{W}^\mathsf{T}, \quad \text{where} \quad \mathbf{W} := \begin{bmatrix} \mathbf{M}^{\frac{1}{2}}\mathbf{V} & \\ & \mathbf{N}^{\frac{1}{2}}\mathbf{U} \end{bmatrix} \quad \text{and} \quad \mathring{\mathbf{K}} := \begin{bmatrix} \mathbf{I} & \Sigma \\ \Sigma^\mathsf{T} & -\mathbf{I} \end{bmatrix}. \tag{3.5}$$

Note that (3.5) is a congruence relation, not a similarity relation, but it expresses any SQD matrix in terms of a related, far simpler SQD matrix that we denote $\mathring{\mathbf{K}}$ and term *canonical*.

Applying the inverse formula (3.3), we see that the leading block of $\mathring{\mathbf{K}}^{-1}$ is $(\mathbf{I} + \Sigma\Sigma^\mathsf{T})^{-1}$, which is a contraction, i.e., has eigenvalues $1/(1+\sigma_i^2) \in (0,1]$. Thus, the leading block of $\mathbf{K}^{-1}$ is a contraction in the $\mathbf{M}$-norm.

The *perfect shuffle* is the permutation

$$\pi := (1, n+1, 2, n+2, \dots, \min(m,n), n+\min(m,n), \dots, n+m),$$

where n is the size of $\mathbf{M}$ and m is the size of $\mathbf{N}$. The permutation π symmetrically permutes $\mathring{\mathbf{K}}$ to block-diagonal form with blocks of size 2 given by

$$\begin{bmatrix} 1 & \sigma_i \\ \sigma_i & -1 \end{bmatrix},$$

except perhaps for the last block, which is $\mathbf{I}$ if $n > m$ and $-\mathbf{I}$ if $n < m$. The block-diagonal form is also tridiagonal, and we will see in the next chapters that it is related to the tridiagonal obtained by applying the Lanczos process to $\mathbf{K}$ with preconditioner blkdiag$(\mathbf{M},\mathbf{N})$.

It is instructive to contrast the above with the symmetric saddle-point matrix

$$\begin{aligned} \mathbf{K}_0 := \begin{bmatrix} \mathbf{M} & \mathbf{A} \\ \mathbf{A}^{\mathsf{T}} & \end{bmatrix} &= \begin{bmatrix} \mathbf{M}^{\frac{1}{2}} & \\ & \mathbf{N}^{\frac{1}{2}} \end{bmatrix} \begin{bmatrix} \mathbf{I} & \overline{\mathbf{A}} \\ \overline{\mathbf{A}}^{\mathsf{T}} & \end{bmatrix} \begin{bmatrix} \mathbf{M}^{\frac{1}{2}} & \\ & \mathbf{N}^{\frac{1}{2}} \end{bmatrix} \\ &= \begin{bmatrix} \mathbf{M}^{\frac{1}{2}}\mathbf{V} & \\ & \mathbf{N}^{\frac{1}{2}}\mathbf{U} \end{bmatrix} \begin{bmatrix} \mathbf{I} & \Sigma \\ \Sigma^{\mathsf{T}} & \end{bmatrix} \begin{bmatrix} \mathbf{V}^{\mathsf{T}}\mathbf{M}^{\frac{1}{2}} & \\ & \mathbf{U}^{\mathsf{T}}\mathbf{N}^{\frac{1}{2}} \end{bmatrix} \\ &= \mathbf{W}\mathring{\mathbf{K}}_0\mathbf{W}^{\mathsf{T}}. \end{aligned}$$

In this case, it is necessary to impose $n > m$ to ensure nonsingularity. For saddle-point matrices, it can be verified that (3.4) remains valid. The leading block of the inverse of the canonical saddle-point matrix, denoted here $\mathring{\mathbf{K}}_0$, may thus be written as

$$\mathbf{I} - \Sigma(\Sigma^{\mathsf{T}}\Sigma)^{-1}\Sigma,$$

and the latter is a projector.

The perfect shuffle again permutes $\mathring{\mathbf{K}}_0$ to block-diagonal/tridiagonal form with blocks

$$\begin{bmatrix} 1 & \sigma_i \\ \sigma_i & \end{bmatrix}$$

followed by a block of zeros.

3.2 ▪ Strong Factorizability

In the second part of his proof, Vanderbei (1995, Theorem 2.1) remarks that all principal submatrices of $\mathbf{P}^{\mathsf{T}}\mathbf{K}\mathbf{P}$ are symmetric permutations of principal submatrices of $\mathbf{K}$. Thus, they are themselves SQD and hence nonsingular. The latter property is sufficient to guarantee existence of the factorization (3.2).

A second proof of (3.2) proposed by Gill, Saunders, and Shinnerl (1996) rests on the factorizability of unsymmetric positive-definite matrices. Observe that $\mathbf{K} = \widehat{\mathbf{K}}\widehat{\mathbf{I}}$, where

$$\widehat{\mathbf{I}} := \begin{bmatrix} \mathbf{I} & \\ & -\mathbf{I} \end{bmatrix} \quad \text{and} \quad \widehat{\mathbf{K}} := \begin{bmatrix} \mathbf{M} & -\mathbf{A} \\ \mathbf{A}^{\mathsf{T}} & \mathbf{N} \end{bmatrix}.$$

$\widehat{\mathbf{K}}$ is unsymmetric but nevertheless positive definite; i.e., its symmetric part

$$\begin{bmatrix} \mathbf{M} & \\ & \mathbf{N} \end{bmatrix}$$

is positive definite. Therefore, for any permutation matrix $\mathbf{P}$,

$$\mathbf{P}^{\mathsf{T}}\mathbf{K}\mathbf{P} = \mathbf{P}^{\mathsf{T}}\widehat{\mathbf{K}}\widehat{\mathbf{I}}\mathbf{P} = \left(\mathbf{P}^{\mathsf{T}}\widehat{\mathbf{K}}\mathbf{P}\right)\left(\mathbf{P}^{\mathsf{T}}\widehat{\mathbf{I}}\mathbf{P}\right). \tag{3.6}$$

The left- and right-hand sides of the above are related according the following result.

Theorem 3.1 (Strong Factorizability).

$$\mathbf{P}^{\mathsf{T}}\mathbf{K}\mathbf{P} = \mathbf{L}\mathbf{D}\mathbf{L}^{\mathsf{T}} \qquad \textit{if and only if} \qquad \mathbf{P}^{\mathsf{T}}\widehat{\mathbf{K}}\mathbf{P} = \mathbf{L}\widetilde{\mathbf{D}}\mathbf{U}, \tag{3.7}$$

where $\widetilde{\mathbf{D}} = \mathbf{D}\widetilde{\mathbf{I}}$ *and* $\widetilde{\mathbf{I}} = \mathbf{P}^{\mathsf{T}}\widehat{\mathbf{I}}\mathbf{P}$ *and where* $\mathbf{U} = \widetilde{\mathbf{I}}\mathbf{L}^{\mathsf{T}}\widetilde{\mathbf{I}}$ *is unit upper triangular.*

Because $\widehat{\mathbf{K}}$ is positive definite, so are also all its principal submatrices, which are therefore nonsingular. That is exactly the condition required by Golub and Van Loan (2013, Theorem 4.1.1) to ensure existence of the $\mathbf{L}\widetilde{\mathbf{D}}\mathbf{U}$ factorization. Taking Theorem 3.1 and the results presented in Golub and Van Loan (2013, Chapter 4, p. 159) into account, we have an alternative proof of the existence of an LU factorization of every symmetric permutation of $\mathbf{K}$. Moreover, because $\widehat{\mathbf{I}}$ only changes signs, (3.7) holds in finite-precision arithmetic.

3.3 ▪ Stability and Pivoting

The following backward-stability result gives general bounds on the error in the finite-precision computation of the factorization of a symmetric permutation of $\mathbf{K}$. It is a combination of results of Golub and Van Loan (1979) and Gill, Saunders, and Shinnerl (1996).

Theorem 3.2 (Stability of SQD Factorization). *Let* $\widetilde{\mathbf{L}}$ *and* $\widetilde{\mathbf{D}}$ *be the computed factors in the finite-precision computation of the factorization* (3.2) *of* $\mathbf{P}^{\mathsf{T}}\mathbf{K}\mathbf{P}$. *Let* u *be the unit roundoff, and let* $\mathbf{z}$ *be the computed solution of* $\mathbf{K}\mathbf{z} = \mathbf{f}$ *obtained by using the standard forward and backward substitution (Golub and Van Loan, 2013, pp. 106–107). Then there exists a perturbation matrix* $\Delta\mathbf{K}$ *such that*

$$(\mathbf{K} + \Delta\mathbf{K})\mathbf{z} = \mathbf{f}$$

with

$$|\Delta\mathbf{K}| \le (m+n)u\left(3|\mathbf{K}| + 5|\widetilde{\mathbf{L}}|\,|\widetilde{\mathbf{D}}|\,|\widetilde{\mathbf{L}}^{\mathsf{T}}|\right) + \mathcal{O}(u^2) \tag{3.8}$$

and

$$\|\Delta\mathbf{K}\|_2 \le u\,\left(3(m+n)\|\mathbf{K}\|_F + 5(m+n)^2\omega(\mathbf{K})\right) + \mathcal{O}(u^2),$$

where

$$\omega(\mathbf{K}) = \left(1 + \frac{\max\left(\|\mathbf{A}^{\mathsf{T}}\mathbf{M}^{-1}\mathbf{A}\|_2,\, \|\mathbf{A}\mathbf{N}^{-1}\mathbf{A}^{\mathsf{T}}\|_2\right)}{\|\mathbf{K}\|_2}\right) \tag{3.9}$$

and where the notation $|\Delta\mathbf{K}|$ *denotes the matrix of absolute values of the elements of* $\Delta\mathbf{K}$.

We already know that when $\mathbf{K}$ is sparse, we can focus on finding a permutation that reduces the fill-in in $\mathbf{L}$. However, as Theorem 3.2 states, we cannot guarantee that the factorization is stable.

Example 3.3. The following toy example given by Golub and Van Loan (1979) illustrates how different permutations can produce unstable factors. Let

$$\mathbf{K} := \begin{bmatrix} 1 & 1 \\ 1 & -\mu \end{bmatrix} = \begin{bmatrix} 1 & 1 \\ 1 & -\mu \end{bmatrix} = \begin{bmatrix} 1 & 0 \\ 1 & 1 \end{bmatrix} \begin{bmatrix} 1 & 0 \\ 0 & -(1+\mu) \end{bmatrix} \begin{bmatrix} 1 & 1 \\ 0 & 1 \end{bmatrix}.$$

All factors are suitably bounded in absolute value, and a numerical factorization can be computed accurately. However, the symmetric permutation

$$\begin{bmatrix} -\mu & 1 \\ 1 & 1 \end{bmatrix} = \begin{bmatrix} 1 & 0 \\ -\frac{1}{\mu} & 1 \end{bmatrix} \begin{bmatrix} -1 & 0 \\ 0 & 1+\frac{1}{\mu} \end{bmatrix} \begin{bmatrix} 1 & -\frac{1}{\mu} \\ 0 & 1 \end{bmatrix} \tag{3.10}$$

yields unbounded factors when $|\mu| \to 0$. ∎

Example 3.3 shows that (3.9) can realistically be very large.

Among other interesting results, George and Ikramov (2000) establish that for any principal submatrix $\mathbf{S}$ of $\mathbf{K}$, we have

$$\operatorname{cond}(\mathbf{K}/\mathbf{S}) \le (1 + \omega(\mathbf{K})) \operatorname{cond}(\mathbf{K}).$$

We now consider pivot strategies based on the symmetric factorization of $\mathbf{K}$ as in (3.2). Consider the symmetric saddle-point system

$$\begin{bmatrix} \mathbf{M} & \mathbf{A} \\ \mathbf{A}^{\mathsf{T}} & 0 \end{bmatrix} \begin{bmatrix} \mathbf{x} \\ \mathbf{y} \end{bmatrix} = \begin{bmatrix} \mathbf{b} \\ 0 \end{bmatrix}. \tag{3.11}$$

By way of a simple regularization $\mathbf{N} = \mu\mathbf{I}$, where $\mu > 0$, we obtain the SQD system

$$\begin{bmatrix} \mathbf{M} & \mathbf{A} \\ \mathbf{A}^{\mathsf{T}} & -\mu\mathbf{I} \end{bmatrix} \begin{bmatrix} \mathbf{x}_\mu \\ \mathbf{y}_\mu \end{bmatrix} = \begin{bmatrix} \mathbf{b} \\ 0 \end{bmatrix}.$$

By the implicit function theorem, as $\mu \to 0$, we can expect that

$$\begin{bmatrix} \mathbf{x}_\mu \\ \mathbf{y}_\mu \end{bmatrix} \to \begin{bmatrix} \mathbf{x} \\ \mathbf{y} \end{bmatrix},$$

provided the limiting matrix is nonsingular. However, Theorem 3.2 implies that for small values of μ, dangerous instabilities could creep in. There are at least two main possibilities of avoiding such instabilities.

A first possibility is to abandon the 1×1 pivot strategy and use instead a 1×1 and 2×2 pivot strategy as described in Golub and Van Loan (2013) and implemented in `MA57` (Harwell Subroutine Library, 2000) by Duff (2004) in a sparse matrix context—see also Duff, Erisman, and Reid (2017).

This option implies relaxing our choice of ordering and accepting to increase the fill-in when the potential instability becomes too high. In addition, $\widetilde{\mathbf{D}}$ becomes block diagonal with blocks of orders 1 and 2.

An implementation of this option begins by introducing a threshold parameter $0 < \rho < 1$. Assume that after $i-1$ steps of Gaussian elimination, we have successfully factorized the the first $i-1$ rows and columns of $\widetilde{\mathbf{K}} = \mathbf{P}\mathbf{K}\mathbf{P}^{\mathsf{T}}$, where $\mathbf{P}$ is a permutation that gives an acceptable level of fill-in. Thus, we have the partition of

$$\widetilde{\mathbf{K}} = \begin{bmatrix} \widetilde{\mathbf{K}}_{11} & \widetilde{\mathbf{K}}_{12} \\ \widetilde{\mathbf{K}}_{12}^{\mathsf{T}} & \widetilde{\mathbf{K}}_{22} \end{bmatrix}, \quad \text{where} \quad \widetilde{\mathbf{K}}_{11} = \widetilde{\mathbf{L}}_1 \widetilde{\mathbf{D}}_1 \widetilde{\mathbf{L}}_1^{\mathsf{T}}.$$

The next step consists in computing the Schur complement

$$\widetilde{\mathbf{K}}/\widetilde{\mathbf{K}}_{11} = \widetilde{\mathbf{K}}_{22} - \widetilde{\mathbf{K}}_{12}^{\mathrm{T}}\widetilde{\mathbf{L}}_1^{-T}\widetilde{\mathbf{D}}_1^{-1}\widetilde{\mathbf{L}}_1^{-1}\widetilde{\mathbf{K}}_{12}. \tag{3.12}$$

Let us denote the candidate 1×1 pivot by $\tilde{\mathrm{k}}_{ii}$, the first entry on the diagonal of (3.12), and let

$$\mathbf{G} = \begin{bmatrix} \tilde{\mathrm{k}}_{i,i} & \tilde{\mathrm{k}}_{i,i+1} \\ \tilde{\mathrm{k}}_{i+1,i} & \tilde{\mathrm{k}}_{i+1,i+1} \end{bmatrix}$$

so that

$$\widetilde{\mathbf{K}}/\widetilde{\mathbf{K}}_{11} = \begin{bmatrix} \tilde{\mathrm{k}}_{i,i} & \tilde{\mathrm{k}}_{i,i+1} & \mathbf{k}_1^{\mathrm{T}} \\ \tilde{\mathrm{k}}_{i+1,i} & \tilde{\mathrm{k}}_{i+1,i+1} & \mathbf{k}_2^{\mathrm{T}} \\ \mathbf{k}_1 & \mathbf{k}_2 & \widetilde{\mathbf{K}}' \end{bmatrix}.$$

Then we define

$$\phi := \frac{\max_{j \neq i+1} |\tilde{\mathrm{k}}_{i,j}|}{|\tilde{\mathrm{k}}_{i,i}|}.$$

In this approach, a pivot $\tilde{\mathrm{k}}_{i,i}$ is considered to be stable if and only if

$$\phi \leq 1/\rho. \tag{3.13}$$

If the test (3.13) fails, the 2×2 matrix $\mathbf{G}$ gives a stable alternative if and only if

$$\psi \leq 1/\rho, \tag{3.14}$$

where

$$\psi := \left\| |\mathbf{G}^{-1}| \begin{bmatrix} \max_{j>i} |\tilde{\mathrm{k}}_{i,j}| \\ \max_{j>i+1} |\tilde{\mathrm{k}}_{i+1,j}| \end{bmatrix} \right\|_\infty$$

and where the max operators scan the rows of $\widetilde{\mathbf{K}}/\widetilde{\mathbf{K}}_{11}$.

In both cases, we can update the rows and the columns using the Schur complement technique (3.12). If both (3.13) and (3.14) fail, the pivot $\mathrm{k}_{i,j}$ is delayed; i.e., we symmetrically permute $\widetilde{\mathbf{K}}/\widetilde{\mathbf{K}}_{11}$ so that the ith diagonal entry appears last. If $\mathbf{K}$ has full rank, we may be able to accept the delayed pivots at a later stage of the Gaussian elimination. We must stress that this could cause the fill-in to increase.

A second possibility is to use some variant of the strategy proposed by Arioli et al. (2007) and Arioli and Duff (2008/2009) in order to avoid delaying pivots. Given a pivot strategy, the *growth factor* is defined as

$$g := \frac{\max\left\{|\tilde{\ell}_{i,j}|, |\tilde{d}_{i,j}| \mid i,j = 1,\ldots,n+m\right\}}{\max\left\{|\tilde{k}_{i,j}| \mid i,j = 1,\ldots,n+m\right\}}.$$

First, if we choose $\rho = \sqrt{u}\|\mathbf{K}\|$, where u is the unit roundoff, we can eliminate a pivot and ensure a growth factor below $1+1/\rho$ if (3.13) and (3.14) are both satisfied, in which case we select the one with the lower growth factor. Finally, if neither (3.13) nor (3.14) is satisfied, we can avoid delaying pivots by perturbing $\tilde{\mathrm{k}}_{i,i}$ using $\delta = \mathrm{sign}(\tilde{\mathrm{k}}_{i,i})(\tau\,|\tilde{\mathrm{k}}_{i,i}|)$, where $\tau = \rho \max_{ij} |\mathrm{k}_{i,j}|$.

We observe that if we choose $\delta = \mathrm{sign}((\mathbf{P}^\mathsf{T}\widehat{\mathbf{I}}\mathbf{P})_{i,i})(\tau|\tilde{\mathrm{k}}_{i,i}|)$, we have dynamically corrected the diagonals of $\mathbf{M}$ and $\mathbf{N}$ in order to obtain a better-regularized problem. Arioli et al. (2007) and Arioli and Duff (2008/2009) establish that despite the potential for a large growth factor, the resulting factorization can be used as preconditioner for the right-preconditioned Flexible GMRES method (Saad, 1993; Duff, 2004). More precisely, the computed factors $\widetilde{\mathbf{L}}$ and $\widetilde{\mathbf{D}}$ satisfy

$$\mathbf{K} + \delta\mathbf{K} + \tau\Delta = \widetilde{\mathbf{L}}\widetilde{\mathbf{D}}\widetilde{\mathbf{L}}^\mathsf{T} := \mathbf{C}, \tag{3.15}$$

$$||\delta\mathbf{K}|| \le c(m+n)u\, ||\, |\widetilde{\mathbf{L}}|\, |\widetilde{\mathbf{D}}|\, |\widetilde{\mathbf{L}}^\mathsf{T}|\, ||, \tag{3.16}$$

$$|\Delta| \le \mathbf{I}. \tag{3.17}$$

Therefore, taking into account that the growth factor in $\widetilde{\mathbf{L}}$ is less than or equal to $1 + 1/\rho$, we can safely assume that

$$c(m+n)\mathbf{u} \left\| \, |\widetilde{\mathbf{L}}|\, |\widetilde{\mathbf{D}}|\, |\widetilde{\mathbf{L}}^\mathsf{T}| \, \right\| \le \tau. \tag{3.18}$$

The right-preconditioned FGMRES for solving $\mathbf{Kz} = \mathbf{b}$ can be stated as Algorithm 3.1.

Algorithm 3.1 FGMRES with Right Preconditioning

$\mathbf{z}_0 = \mathbf{C}^{-1}\mathbf{b}$, $\mathbf{r}_0 = \mathbf{b} - \mathbf{K}\mathbf{z}_0$ and $\beta = ||\mathbf{r}_0||$
$\mathbf{v}_1 = \mathbf{r}_0/\beta$; $k = 0$
while $||\mathbf{r}_k|| > \mathbf{u}(||\mathbf{b}|| + ||\mathbf{K}||\,||\mathbf{z}_k||)$ **do**
  $k = k+1$, $\mathbf{z}_k = \mathbf{C}^{-1}\mathbf{v}_k$, and $\mathbf{w} = \mathbf{K}\mathbf{z}_k$
  for $i = 1, \ldots, k$ **do**
    $h_{i,k} = \mathbf{v}_i^\mathsf{T}\mathbf{w}$; $\mathbf{w} = \mathbf{w} - h_{i,k}\mathbf{v}_i$;
  end for
  $h_{k+1,k} = ||\mathbf{w}||$; $\mathbf{v}_{k+1} = \mathbf{w}/h_{k+1,k}$;
  $\mathbf{Z}_k = [\mathbf{z}_1, \ldots, \mathbf{z}_k]$; $\mathbf{V}_k = [\mathbf{v}_1, \ldots, \mathbf{v}_k]$;
  $\mathbf{H}_k = \{h_{i,j}\}_{1 \le i \le j+1; 1 \le j \le k}$;
  $\mathbf{y}_k = \arg\min_\mathbf{y} ||\beta\mathbf{e}_1 - \mathbf{H}_k\mathbf{y}||$;
  $\mathbf{z}_k = \mathbf{z}_0 + \mathbf{Z}_k\mathbf{y}_k$ and $\mathbf{r}_k = \mathbf{b} - \mathbf{K}\mathbf{z}_k$;
end while

Under the assumption (3.18) and assuming that the upper Hessenberg matrices $\mathbf{H}_k$ are not too ill conditioned, the Algorithm 3.1 will compute a normwise backward stable solution (Arioli et al., 2007, Theorems 5.1, 5.2, and 5.3). Arioli and Duff (2008/2009) prove that it is possible to obtain high accuracy in the solution by using mixed-precision IEEE arithmetic.

3.4 ▪ Iterative Strategies

The iterative methods that come to mind to solve a SQD system are MINRES and SYMMLQ. The question arises as to how to precondition the iterations. In this section, we review a few preconditioning strategies found in the literature and an additional iterative method resembling the conjugate gradient method.

3.4.1 ▪ Constraint preconditioners

Constraint preconditioners form a class of popular and effective preconditioners for both symmetric saddle-point systems (Keller, Gould, and Wathen, 2000) and regularized symmetric saddle-point systems (Dollar et al., 2006). Even though they apply to systems with an indefinite leading block, we restrict our attention to the case of a positive-definite leading block. A constraint preconditioner for (2.1) has the form

$$\begin{bmatrix} \widetilde{\mathbf{M}} & \mathbf{A} \\ \mathbf{A}^\mathsf{T} & 0 \end{bmatrix}, \tag{3.19}$$

where $\widetilde{\mathbf{M}}$ is a symmetric and positive-definite approximation of $\mathbf{M}$ chosen in such a way that solving systems with coefficient (3.19) is computationally realistic. The quality of a constraint preconditioner is governed by the quality of $\widetilde{\mathbf{M}}$ as an approximation of $\mathbf{M}$ on the nullspace of $\mathbf{A}^\mathsf{T}$ (Keller et al., 2000, Theorem 2.1). Preconditioning systems are typically solved by way of a one-time symmetric indefinite factorization and multiple forward and backsolves. However, such a factorization is realistic only if $\mathbf{A}$ and the factors of (3.19) are sufficiently sparse. Gould, Orban, and Rees (2013) establish that constraint-preconditioned Krylov methods are equivalent in exact arithmetic to projected Krylov methods provided that the right-hand side has a zero bottom block as in (2.1), and they demonstrate how to derive projected and constraint-preconditioned variants of any Krylov or Krylov-type method. In particular, the authors experiment with constraint-preconditioned and projected variants of MINRES, but the strategy applies even when $\mathbf{M}$ is square and unsymmetric.

As the general transformation (1.7) suggests, strategies developed for standard saddle-point systems may be converted into strategies for SQD systems. Dollar et al. (2006) develop a constraint-preconditioned variant of the conjugate gradient method for

$$\begin{bmatrix} \mathbf{M} & \mathbf{A} \\ \mathbf{A}^\mathsf{T} & -\mathbf{N} \end{bmatrix} \begin{bmatrix} \mathbf{x} \\ \mathbf{y} \end{bmatrix} = \begin{bmatrix} \mathbf{b} \\ 0 \end{bmatrix}, \tag{3.20}$$

where the preconditioner takes the form

$$\begin{bmatrix} \widetilde{\mathbf{M}} & \mathbf{A} \\ \mathbf{A}^\mathsf{T} & -\mathbf{N} \end{bmatrix}, \tag{3.21}$$

where, as above, $\widetilde{\mathbf{M}}$ is a suitable symmetric approximation of $\mathbf{M}$. The requirements on $\widetilde{\mathbf{M}}$, which can be deduced immediately from (1.7), are that

$$\begin{bmatrix} \widetilde{\mathbf{M}} & \\ & \mathbf{D}^{-1} \end{bmatrix}$$

should be positive definite on the nullspace of

$$\begin{bmatrix} \mathbf{A}^\mathsf{T} & \mathbf{L} \end{bmatrix}.$$

In the SQD context where $\mathbf{M}$ and $\mathbf{D}$ are positive definite, it seems natural to select $\widetilde{\mathbf{M}}$ also positive definite, but the methodology of Dollar et al. (2006) also applies when $\mathbf{M}$ and $\mathbf{N}$ are indefinite or singular.

3.4.2 ▪ Incomplete factorization preconditioners

Lin and Moré (1999) propose an incomplete Cholesky factorization scheme for symmetric matrices that allows the user to control the amount of memory reserved for the incomplete factors. More precisely, if $\mathbf{H}$ is $n \times n$ and symmetric but not necessarily positive definite with q nonzero elements in its lower (or upper) triangle, the user selects an integer p, and the procedure computes a shift $\alpha \geq 0$ and triangular matrices $\widetilde{\mathbf{L}}$ such that

$$\mathbf{H} + \alpha \mathbf{I} = \widetilde{\mathbf{L}}\widetilde{\mathbf{L}}^{\mathsf{T}} + \mathbf{E},$$

where $\mathbf{H} + \alpha \mathbf{I}$ is positive definite, $\mathbf{L}$ has at most $q + pn$ nonzero elements, and $\mathbf{E}$ is the error. When such a factorization exists, $\widetilde{\mathbf{L}}$ can be used as a preconditioner for systems involving $\mathbf{H}$. Lin and Moré (1999) establish that the incomplete Cholesky factorization exists provided that $\mathbf{H}$ is an H-matrix, which is a condition similar to but weaker than strict diagonal dominance (Axelsson, 1994).

Orban (2015) generalizes the ideas of Lin and Moré (1999) to block matrices in hopes of computing preconditioners for SQD systems by way of incomplete $\mathbf{LDL}^{\mathsf{T}}$ factorizations. Suppose $\mathbf{K}$ in (3.1) is symmetric but not necessarily SQD, the user selects an integer p, and the procedure computes a unit triangular matrix $\widetilde{\mathbf{L}}$, a diagonal matrix $\widetilde{\mathbf{D}}$, and a shift $\alpha > 0$ such that

$$\begin{bmatrix} \mathbf{M} + \alpha \mathbf{I} & \mathbf{A} \\ \mathbf{A}^{\mathsf{T}} & -(\mathbf{N} + \alpha \mathbf{I}) \end{bmatrix} = \widetilde{\mathbf{L}}\widetilde{\mathbf{D}}\widetilde{\mathbf{L}}^{\mathsf{T}} + \mathbf{E},$$

where $\mathbf{L}$ has at most $q + p(n + m)$ nonzeros and q is the number of nonzeros in the lower triangle of $\mathbf{K}$. The intuition is that even if $\mathbf{M}$ and/or $\mathbf{N}$ is not positive definite, there exist sufficiently large shifts such that the shifted matrix is SQD. As above, it can be shown that the incomplete factorization exists, provided that $\mathbf{K}$ is an H-matrix with no zeros on the diagonal, but that is only a sufficient condition, and in our experience, the incomplete factorization exists before the shifts are large enough to obtain an H-matrix.

When $\mathbf{K} = \mathbf{LDL}^{\mathsf{T}}$ is SQD, $\mathbf{L}|\mathbf{D}|\mathbf{L}^{\mathsf{T}}$ is an effective positive-definite preconditioner, where $|\mathbf{D}|$ is the diagonal matrix made of the absolute values of the elements on the diagonal of $\mathbf{D}$. Indeed, the preconditioned matrix

$$|\mathbf{D}|^{-\frac{1}{2}} \mathbf{L}^{-1} \mathbf{K} \mathbf{L}^{-\mathsf{T}} |\mathbf{D}|^{-\frac{1}{2}}$$

possesses only the two distinct eigenvalues ± 1. The idea then is to compute incomplete factors $\widetilde{\mathbf{L}}$ and $\widetilde{\mathbf{D}}$ and use $\widetilde{\mathbf{L}}|\widetilde{\mathbf{D}}|\widetilde{\mathbf{L}}^{\mathsf{T}}$ as preconditioner for systems involving $\mathbf{K}$, for example in the preconditioned MINRES. When $\mathbf{K}$ is already sufficiently SQD, meaning that its smallest eigenvalues in absolute value are safely away from zero, the shift α will remain zero. In other cases, for example when $\mathbf{K}$ is symmetric but not SQD, a nonzero shift will be computed. The shift α will be added to diagonal elements that were nonnegative in $\mathbf{K}$ and subtracted from diagonal elements that were negative in $\mathbf{K}$. Orban (2015) shows extensive numerical results that illustrate the effectiveness of such an incomplete factorization approach on a collection of SQD systems arising in interior-point methods for convex quadratic optimization.

Scott and Tůma (2014) compute a similar limited-memory factorization using additional storage during the factorization phase.

3.4.3 ▪ Preconditioners for least-squares problems

Arioli and Duff (2015) propose an approach to solve the least-squares problem

$$\underset{\mathbf{x}\in\mathbb{R}^n}{\text{minimize}} \ \|\mathbf{Ax}-\mathbf{b}\|$$

that partitions

$$\mathbf{A}=\begin{bmatrix}\mathbf{A}_1\\ \mathbf{A}_2\end{bmatrix},$$

where $\mathbf{A}_1 \in \mathbb{R}^{n\times n}$ is nonsingular. Then the augmented system

$$\begin{bmatrix}\mathbf{I}_n & & \mathbf{A}_1\\ & \mathbf{I}_{m-n} & \mathbf{A}_2\\ \mathbf{A}_1^\mathsf{T} & \mathbf{A}_2^\mathsf{T} & 0\end{bmatrix}\begin{bmatrix}\mathbf{x}_1\\ \mathbf{x}_2\\ \mathbf{y}\end{bmatrix}=\begin{bmatrix}\mathbf{b}_1\\ \mathbf{b}_2\\ 0\end{bmatrix} \tag{3.22}$$

reduces to the SQD system

$$\begin{bmatrix}\mathbf{I}_{m-n} & \mathbf{A}_2\\ \mathbf{A}_2^\mathsf{T} & -\mathbf{A}_1^\mathsf{T}\mathbf{A}_1\end{bmatrix}\begin{bmatrix}\mathbf{x}_2\\ \mathbf{y}\end{bmatrix}=\begin{bmatrix}\mathbf{b}_2\\ -\mathbf{A}_1^\mathsf{T}\mathbf{b}_1\end{bmatrix}. \tag{3.23}$$

The latter SQD system can be solved efficiently by any of the algorithms described in subsequent chapters, provided that systems with coefficient $\mathbf{A}_1^\mathsf{T}\mathbf{A}_1$ can be solved efficiently. Arioli and Duff (2015) give numerical evidence that (3.23) has favorable eigenvalue distribution for several classes of problems.

3.4.4 ▪ Lanczos method for unsymmetric positive-definite systems

Earlier, we discussed how (3.1) can be transformed by the simple scaling

$$\widehat{\mathbf{I}}=\begin{bmatrix}\mathbf{I} & \\ & -\mathbf{I}\end{bmatrix}$$

into the matrix

$$\widehat{\mathbf{K}}=\begin{bmatrix}\mathbf{M} & -\mathbf{A}\\ \mathbf{A}^\mathsf{T} & \mathbf{N}\end{bmatrix}=\mathbf{K}\widehat{\mathbf{I}}$$

that is now unsymmetric but nevertheless positive definite; i.e., its symmetric part

$$\mathbf{H}=\tfrac{1}{2}\left(\widehat{\mathbf{K}}+\widehat{\mathbf{K}}^\mathsf{T}\right)=\begin{bmatrix}\mathbf{M} & \\ & \mathbf{N}\end{bmatrix}$$

is SPD. Its skew-symmetric part is

$$\mathbf{S}=\tfrac{1}{2}\left(\widehat{\mathbf{K}}-\widehat{\mathbf{K}}^\mathsf{T}\right)=\begin{bmatrix} & -\mathbf{A}\\ \mathbf{A}^\mathsf{T} & \end{bmatrix}.$$

The general setting is that of a finite-dimensional Hilbert space $\mathbb{H}$ endowed with the inner product $(\mathbf{w},\mathbf{z})_\mathbf{H} := \mathbf{w}^\mathsf{T}\mathbf{Hz}$ and induced norm $\|\mathbf{z}\|_\mathbf{H}^2=\mathbf{z}^\mathsf{T}\mathbf{Hz}$. Concus and Golub (1975), Widlund (1978), and Szyld and Widlund (1993) introduce Lanczos methods specialized to unsymmetric positive-definite systems.

Consider the natural splitting of $\widehat{\mathbf{K}}$

$$\widehat{\mathbf{K}}=\mathbf{H}-\mathbf{S}.$$

Any system $\mathbf{Kw} = \mathbf{h}$ can be converted to $\widehat{\mathbf{K}}\widehat{\mathbf{w}} = \mathbf{h}$. The main idea is to rewrite the latter equivalently as $(\mathbf{I} - \mathbf{H}^{-1}\mathbf{S})\widehat{\mathbf{w}} = \mathbf{H}^{-1}\mathbf{h}$ and to apply a variant of the conjugate-gradient method to $\mathbf{G} := \mathbf{H}^{-1}\mathbf{S}$. Note that $\mathbf{G}$ is skew-symmetric in the $\mathbf{H}$-inner product, which means that for all $\mathbf{w}$ and $\mathbf{z}$ in $\mathbb{H}$,

$$(\mathbf{w}, \mathbf{Gz})_{\mathbf{H}} = \mathbf{w}^{\mathsf{T}}\mathbf{Sz} = -\mathbf{w}^{\mathsf{T}}\mathbf{S}^{\mathsf{T}}\mathbf{z} = (-\mathbf{Gw}, \mathbf{z})_{\mathbf{H}}.$$

Szyld and Widlund (1993) seek approximations of the solutions of

$$\begin{bmatrix} \mathbf{M} & -\mathbf{A} \\ \mathbf{A}^{\mathsf{T}} & \mathbf{N} \end{bmatrix} \begin{bmatrix} \mathbf{x} \\ -\mathbf{y} \end{bmatrix} = \begin{bmatrix} \mathbf{f} \\ \mathbf{g} \end{bmatrix}$$

in the Krylov subspaces

$$\mathscr{K}_i(\mathbf{H}^{-1}\mathbf{S}, \mathbf{v}^{(0)}) = \left\{\mathbf{v}^{(0)}, \mathbf{H}^{-1}\mathbf{S}\mathbf{v}^{(0)}, \ldots, (\mathbf{H}^{-1}\mathbf{S})^i\mathbf{v}^{(0)}\right\}, \qquad i \geq 0,$$

where

$$\mathbf{w}^{(0)} = \begin{bmatrix} \mathbf{w}_1^{(0)} \\ \mathbf{w}_2^{(0)} \end{bmatrix} = \begin{bmatrix} \mathbf{f} - \mathbf{M}^{-1}\mathbf{A}\mathbf{y}^{(0)} \\ \mathbf{g} - \mathbf{N}^{-1}\mathbf{A}^{\mathsf{T}}\mathbf{x}^{(0)} \end{bmatrix}.$$

Without any loss of generality, we can assume that $\mathbf{x}^{(0)} = 0$ and $\mathbf{y}^{(0)} = 0$.

The fact that $\mathbf{G}$ is skew-symmetric in the $\mathbf{H}$-inner product allows us to compute an $\mathbf{H}$-orthogonal basis of $\mathscr{K}_i$ using a three-term recurrence formula. Algorithm 3.2 is a specialized version of the classical conjugate-gradient method.

Algorithm 3.2 CG for Nonsymmetric Positive-Definite Matrices

Require: $\mathbf{x}^{(-1)} = \mathbf{x}^{(0)} = 0$, $\mathbf{y}^{(-1)} = \mathbf{y}^{(0)} = 0$, and $\omega_1 = 1$;

for $i = 0, 1, \ldots$ **do**

  Solve

$$\begin{bmatrix} \mathbf{M} & \\ & \mathbf{N} \end{bmatrix} \begin{bmatrix} \mathbf{w}_1^{(i)} \\ \mathbf{w}_2^{(i)} \end{bmatrix} = \begin{bmatrix} \mathbf{f} - \mathbf{A}\mathbf{y}^{(i)} \\ \mathbf{g} - \mathbf{A}^{\mathsf{T}}\mathbf{x}^{(i)} \end{bmatrix}$$

  Compute $\eta_i = \mathbf{w}_1^{(i)T}\mathbf{M}\mathbf{w}_1^{(i)} + \mathbf{w}_2^{(i)T}\mathbf{N}\mathbf{w}_2^{(i)}$

  if $i \geq 1$ **then**

    Compute

$$\omega_{i+1} = \frac{\omega_i \eta_{i-1}}{\omega_i \eta_{i-1} + \eta_i}$$

  end if

  Compute

$$\begin{bmatrix} \mathbf{x}^{(i+1)} \\ \mathbf{y}^{(i+1)} \end{bmatrix} = \begin{bmatrix} \mathbf{x}^{(i-1)} + \omega_{i+1}\left(\mathbf{w}_1^{(}i) + \mathbf{x}^{(i)} - \mathbf{x}^{(i-1)})\right) \\ \mathbf{y}^{(i-1)} + \omega_{i+1}\left(\mathbf{w}_2^{(i)} + \mathbf{y}^{(i)} - \mathbf{y}^{(i-1)})\right) \end{bmatrix}$$

end for

In exact arithmetic, Algorithm 3.2 converges in a finite number of steps (Widlund, 1978; Szyld and Widlund, 1993). Several numerical experiments presented in Widlund (1978) support the efficiency of this method for problems obtained by the approximation of PDEs.

Chapter 4

Fundamental Processes

4.1 ▪ The Lanczos Process

In what follows, we often refer to Lanczos processes in various contexts. To establish the notation, consider a generic symmetric linear system with matrix $\mathbf{H}$ and right-hand side $\mathbf{d}$. A Lanczos (1950, 1952) process applied to $\mathbf{H}$ and $\mathbf{d}$ constructs a sequence of vectors $\{\mathbf{s}_k\}$ according to the recursion

$$
\begin{aligned}
\omega_1 \mathbf{s}_1 &= \mathbf{d}, \\
\omega_{k+1}\mathbf{s}_{k+1} &= \mathbf{H}\mathbf{s}_k - \chi_k \mathbf{s}_k - \omega_k \mathbf{s}_{k-1}, \qquad \chi_k := \mathbf{s}_k^{\mathsf{T}}\mathbf{H}\mathbf{s}_k,
\end{aligned}
\tag{4.1}
$$

with $\mathbf{s}_0 := 0$ and ω_k chosen at each iteration so that $\|\mathbf{s}_k\|_2 = 1$. The notation in (4.1) is intentionally nonstandard to avoid confusion with scalar and vector quantities defined in the rest of the book, the latter adhering more closely to standard notation found in the literature. It is possible to derive (4.1) from the standard Gram–Schmidt orthogonalization process by exploiting the symmetry of $\mathbf{H}$, and therefore the vectors $\mathbf{s}_k$ are theoretically orthonormal. After k iterations, the Lanczos process has generated $k+1$ vectors. We gather the first k Lanczos vectors into the matrix $\mathbf{S}_k := \begin{bmatrix}\mathbf{s}_1 & \cdots & \mathbf{s}_k\end{bmatrix}$. The situation after k iterations may be summarized as

$$
\mathbf{H}\mathbf{S}_k = \mathbf{S}_k \mathbf{\Omega}_k + \omega_{k+1}\mathbf{s}_{k+1}\mathbf{e}_k^{\mathsf{T}},
\tag{4.2}
$$

where $\mathbf{\Omega}_k$ is the tridiagonal matrix

$$
\mathbf{\Omega}_k := \begin{bmatrix} \chi_1 & \omega_2 & & \\ \omega_2 & \chi_2 & \ddots & \\ & \ddots & \ddots & \omega_k \\ & & \omega_k & \chi_k \end{bmatrix}.
$$

In floating-point arithmetic, orthogonality of the vectors $\mathbf{s}_k$ is soon compromised, and

$$
\mathbf{S}_k^{\mathsf{T}}\mathbf{H}\mathbf{S}_k = \mathbf{\Omega}_k
$$

typically holds only in exact arithmetic. By contrast, (4.2) generally holds to within machine precision.

4.2 ▪ The Generalized Golub–Kahan Bidiagonalization Process

Motivated by the least-squares problems of the previous chapter, we now describe the standard Golub and Kahan (1965) bidiagonalization process that forms the basis of several numerical methods for such problems. The standard Golub–Kahan bidiagonalization process with initial vector $\mathbf{b} \in \mathbb{R}^n$ can be stated as Algorithm 4.1.

Algorithm 4.1 Golub–Kahan Bidiagonalization

Require: $\overline{\mathbf{A}}, \overline{\mathbf{b}}$

$\beta_1 \overline{\mathbf{u}}_1 = \overline{\mathbf{b}}$ with $\beta_1 > 0$ so that $\|\overline{\mathbf{u}}_1\|_2 = 1$

$\alpha_1 \overline{\mathbf{v}}_1 = \overline{\mathbf{A}}^\mathsf{T} \overline{\mathbf{u}}_1$ with $\alpha_1 > 0$ so that $\|\overline{\mathbf{v}}_1\|_2 = 1$

for $k = 1, 2, \ldots$ **do**

  $\beta_{k+1} \overline{\mathbf{u}}_{k+1} = \overline{\mathbf{A}} \overline{\mathbf{v}}_k - \alpha_k \overline{\mathbf{u}}_k$ with $\beta_{k+1} > 0$ so that $\|\overline{\mathbf{u}}_{k+1}\|_2 = 1$

  $\alpha_{k+1} \overline{\mathbf{v}}_{k+1} = \overline{\mathbf{A}}^\mathsf{T} \overline{\mathbf{u}}_{k+1} - \beta_{k+1} \overline{\mathbf{v}}_k$ with $\alpha_{k+1} > 0$ so that $\|\overline{\mathbf{v}}_{k+1}\|_2 = 1$.

end for

In exact arithmetic, Algorithm 4.1 generates two sets of orthonormal vectors, $\{\overline{\mathbf{u}}_i\}$ and $\{\overline{\mathbf{v}}_i\}$, that can be used to determine the left and right singular vectors of $\overline{\mathbf{A}}$ (Golub and Kahan, 1965).

Consider the application of Algorithm 4.1 to the operator $\overline{\mathbf{A}} = \mathbf{M}^{-\frac{1}{2}} \mathbf{A} \mathbf{N}^{-\frac{1}{2}}$ with initial vector $\overline{\mathbf{b}} = \mathbf{M}^{-\frac{1}{2}} \mathbf{b}$. It is straightforward to verify that after the change of variable $\mathbf{u}_i := \mathbf{M}^{-\frac{1}{2}} \overline{\mathbf{u}}_i$ and $\mathbf{v}_i := \mathbf{N}^{-\frac{1}{2}} \overline{\mathbf{v}}_i$, the resulting process may be written as Algorithm 4.2.

Algorithm 4.2 Generalized Golub–Kahan Bidiagonalization, first variant

---

Require: $\mathbf{M}, \mathbf{A}, \mathbf{N}, \mathbf{b}$

$\beta_1 \mathbf{M} \mathbf{u}_1 = \mathbf{b}$ with $\beta_1 > 0$ so that $\|\mathbf{u}_1\|_\mathbf{M} = 1$

$\alpha_1 \mathbf{N} \mathbf{v}_1 = \mathbf{A}^\mathsf{T} \mathbf{u}_1$ with $\alpha_1 > 0$ so that $\|\mathbf{v}_1\|_\mathbf{N} = 1$

for $k = 1, 2, \ldots$ **do**

  $\beta_{k+1} \mathbf{M} \mathbf{u}_{k+1} = \mathbf{A} \mathbf{v}_k - \alpha_k \mathbf{M} \mathbf{u}_k$ with $\beta_{k+1} > 0$ so that $\|\mathbf{u}_{k+1}\|_\mathbf{M} = 1$

  $\alpha_{k+1} \mathbf{N} \mathbf{v}_{k+1} = \mathbf{A}^\mathsf{T} \mathbf{u}_{k+1} - \beta_{k+1} \mathbf{N} \mathbf{v}_k$ with $\alpha_{k+1} > 0$ so that $\|\mathbf{v}_{k+1}\|_\mathbf{N} = 1$.

end for

We refer to Algorithm 4.2 as the *generalized Golub–Kahan bidiagonalization* process in the sense that the left and right singular vectors, $\{\mathbf{u}_i\}$ and $\{\mathbf{v}_i\}$, are orthonormal with respect to inner products defined by $\mathbf{M}$ and $\mathbf{N}$, respectively. It is important to note that at each iteration, one solve with $\mathbf{M}$ and one solve with $\mathbf{N}$ must be performed. Indeed, the terms $\mathbf{M}\mathbf{u}_k$ and $\mathbf{N}\mathbf{v}_k$ in the right-hand sides of the assignments in the loop were computed during the previous pass through the loop. For instance, the computation of β_{k+1} and $\mathbf{u}_{k+1}$ could be detailed as

1. set $\hat{\mathbf{u}}_{k+1} = \mathbf{A}\mathbf{v}_k - \alpha_k (\hat{\mathbf{u}}_k / \beta_k)$;
2. solve $\mathbf{M} \widetilde{\mathbf{u}}_{k+1} = \hat{\mathbf{u}}_{k+1}$ for $\widetilde{\mathbf{u}}_{k+1}$;
3. set $\beta_{k+1} = \sqrt{\widetilde{\mathbf{u}}_{k+1}^\mathsf{T} \hat{\mathbf{u}}_{k+1}}$;
4. set $\mathbf{u}_{k+1} = \widetilde{\mathbf{u}}_{k+1} / \beta_{k+1}$.

The storage per iteration required by Algorithm 4.2 is the same as that required by Algorithm 4.1 with the addition of one n-vector for $\mathbf{Mu}_k$ and one m-vector for $\mathbf{Nv}_k$. The computational effort per iteration is that of Algorithm 4.1 with the addition of one solve with $\mathbf{M}$ and one solve with $\mathbf{N}$.

After k steps of Algorithm 4.2, the situation can be summarized as

$$\mathbf{AV}_k = \mathbf{MU}_k\mathbf{B}_k + \beta_{k+1}\mathbf{Mu}_{k+1}\mathbf{e}_k^\mathsf{T} \tag{4.3a}$$
$$= \mathbf{MU}_{k+1}\mathbf{E}_k, \tag{4.3b}$$
$$\mathbf{A}^\mathsf{T}\mathbf{U}_{k+1} = \mathbf{NV}_{k+1}\mathbf{B}_{k+1}^\mathsf{T} \tag{4.3c}$$
$$= \mathbf{NV}_k\mathbf{E}_k^\mathsf{T} + \alpha_{k+1}\mathbf{Nv}_{k+1}\mathbf{e}_{k+1}^\mathsf{T}, \tag{4.3d}$$

where $\mathbf{e}_k$ is the kth vector of the canonical basis; $\mathbf{U}_k$ and $\mathbf{V}_k$ are the $n \times k$ and $m \times k$ matrices whose columns are $\mathbf{u}_1$ through $\mathbf{u}_k$ and $\mathbf{v}_1$ through $\mathbf{v}_k$, respectively; and

$$\mathbf{B}_k := \begin{bmatrix} \alpha_1 & & & \\ \beta_2 & \alpha_2 & & \\ & \ddots & \ddots & \\ & & \beta_k & \alpha_k \end{bmatrix}, \quad \mathbf{E}_k := \begin{bmatrix} \alpha_1 & & & \\ \beta_2 & \alpha_2 & & \\ & \ddots & \ddots & \\ & & \beta_k & \alpha_k \\ & & & \beta_{k+1} \end{bmatrix} = \begin{bmatrix} \mathbf{B}_k \\ \beta_{k+1}\mathbf{e}_k^\mathsf{T} \end{bmatrix}; \tag{4.4}$$

i.e., $\mathbf{B}_k$ is $k \times k$ lower bidiagonal and $\mathbf{E}_k$ is $\mathbf{B}_k$ with one extra row. The orthogonality properties of the vectors $\mathbf{u}_j$ and $\mathbf{v}_j$ imply that the matrices $\mathbf{M}^{\frac{1}{2}}\mathbf{U}_k$ and $\mathbf{N}^{\frac{1}{2}}\mathbf{V}_k$ are orthogonal for all k.

Algorithm 4.2 is a generalization of the process referred to as GKLB(**M**) by Benbow (1999), and it may be denoted GKLB(**M**, **N**). An analysis mirroring that of Paige (1974) is instructive in relation to the termination of the generalized Golub–Kahan process. The first situation that can cause the process to stop is that $\beta_{k+1} = 0$. In this case, we obtain from (4.3) that, in exact arithmetic,

$$\mathbf{AV}_k = \mathbf{MU}_k\mathbf{B}_k, \quad \mathbf{A}^\mathsf{T}\mathbf{U}_k = \mathbf{NV}_k\mathbf{B}_k^\mathsf{T}, \quad \mathbf{U}_k^\mathsf{T}\mathbf{MU}_k = \mathbf{I}_k, \quad \mathbf{V}_k^\mathsf{T}\mathbf{NV}_k = \mathbf{I}_k. \tag{4.5}$$

The second situation is that $\alpha_{k+1} = 0$. In this case, in exact arithmetic,

$$\mathbf{A}^\mathsf{T}\mathbf{U}_{k+1} = \mathbf{NV}_k\mathbf{E}_k^\mathsf{T}, \quad \mathbf{AV}_k = \mathbf{MU}_{k+1}\mathbf{E}_k, \quad \mathbf{U}_{k+1}^\mathsf{T}\mathbf{MU}_{k+1} = \mathbf{I}_{k+1}, \quad \mathbf{V}_k^\mathsf{T}\mathbf{NV}_k = \mathbf{I}_k. \tag{4.6}$$

Let $\mathbf{E}_k = \mathbf{P}_{k+1}\Sigma_k\mathbf{Q}_k^\mathsf{T}$ be the singular value decomposition of $\mathbf{E}_k$, where $\mathbf{P}_{k+1}$ is orthogonal $(k+1) \times (k+1)$, Σ_k is $(k+1) \times k$, $\mathbf{Q}_k$ orthogonal is $k \times k$, and

$$\Sigma_k = \begin{bmatrix} \sigma_1 & & & \\ & \sigma_2 & & \\ & & \ddots & \\ & & & \sigma_k \\ 0 & 0 & \dots & 0 \end{bmatrix}, \qquad \sigma_i > 0, \quad i = 1, \dots, k.$$

We have from (4.6) that $\mathbf{A}^\mathsf{T}\mathbf{U}_{k+1}\mathbf{P}_{k+1} = \mathbf{NV}_k\mathbf{Q}_k\Sigma_k^\mathsf{T}$ and $\mathbf{AV}_k\mathbf{Q}_k = \mathbf{MU}_{k+1}\mathbf{P}_{k+1}\Sigma_k$. Equivalently, the last two identities can be stated as

$$(\mathbf{M}^{-\frac{1}{2}}\mathbf{AN}^{-\frac{1}{2}})^\mathsf{T}\overline{\mathbf{U}}_{k+1}\mathbf{P}_{k+1} = \overline{\mathbf{V}}_k\mathbf{Q}_k\Sigma_k^\mathsf{T}, \tag{4.7a}$$
$$(\mathbf{M}^{-\frac{1}{2}}\mathbf{AN}^{-\frac{1}{2}})\overline{\mathbf{V}}_k\mathbf{Q}_k = \overline{\mathbf{U}}_{k+1}\mathbf{P}_{k+1}\Sigma_k, \tag{4.7b}$$

where we used $\bar{\mathbf{u}}_j = \mathbf{M}^{\frac{1}{2}}\mathbf{u}_j$ and $\bar{\mathbf{v}}_j = \mathbf{N}^{\frac{1}{2}}\mathbf{v}_j$ for all j. The relations (4.7) show that $\bar{\mathbf{A}} := \mathbf{M}^{-\frac{1}{2}}\mathbf{A}\mathbf{N}^{-\frac{1}{2}}$ has a zero singular value and that the associated right singular vector is the last column of $\bar{\mathbf{U}}_{k+1}\mathbf{P}_{k+1}$. This last column can be written as

$$\mathbf{t} = \sum_{j=1}^{k+1} p_{k+1,j}\mathbf{M}^{\frac{1}{2}}\mathbf{u}_j,$$

where $\mathbf{p}_{k+1}$ is the last column of $\mathbf{P}_{k+1}$. This vector $\mathbf{t}$ is a linear combination of the vectors $\mathbf{M}^{\frac{1}{2}}\mathbf{u}_j$ and must lie in $\text{Null}(\bar{\mathbf{A}}^{\mathsf{T}}) = \text{Null}(\mathbf{A}^{\mathsf{T}}\mathbf{M}^{-\frac{1}{2}})$. Assume now that $\mathbf{b} \in \text{Range}(\mathbf{A})$. Then $\mathbf{u}_1 \in \text{Range}(\mathbf{M}^{-1}\mathbf{A})$. A recursion argument easily establishes that each $\mathbf{u}_j \in \text{Range}(\mathbf{M}^{-1}\mathbf{A})$. In this case, the vector $\mathbf{t}$ thus lies in $\text{Range}(\mathbf{M}^{-\frac{1}{2}}\mathbf{A})$, which is in contradiction with the previous conclusion that $\mathbf{t} \in \text{Null}(\mathbf{A}^{\mathsf{T}}\mathbf{M}^{-\frac{1}{2}})$. Therefore, if $\mathbf{b} \in \text{Range}(\mathbf{A})$, Algorithm 4.2 cannot terminate with $\alpha_{k+1} = 0$. It must thus terminate with $\beta_{k+1} = 0$, and the final situation must be described by (4.5).

Conversely, if (4.5) describes the final situation, necessarily $\mathbf{u}_1 \in \text{Range}(\mathbf{M}^{-1}\mathbf{A})$; i.e., $\mathbf{b} \in \text{Range}(\mathbf{A})$ because $\mathbf{U}_k = \mathbf{M}^{-1}\mathbf{A}\mathbf{V}_k\mathbf{B}_k^{-1}$, which shows that all $\mathbf{u}_j$ lie in $\text{Range}(\mathbf{M}^{-1}\mathbf{A})$. Therefore, if $\mathbf{b} \notin \text{Range}(\mathbf{A})$, the final situation must be described by (4.6).

In the same way that Algorithm 4.1 is closely related to the singular-value decomposition of $\mathbf{A}$, Algorithm 4.2 is related to the elliptic singular-value decomposition of $\mathbf{A}$ (Arioli, 2013) in the sense that the families $\{\mathbf{u}_i\}$ and $\{\mathbf{v}_i\}$ determine families $\{\mathbf{w}_i\}$ and $\{\mathbf{z}_i\}$ satisfying

$$\mathbf{A}\mathbf{z}_i = \sigma_i\mathbf{M}\mathbf{w}_i, \qquad \mathbf{z}_i^{\mathsf{T}}\mathbf{N}\mathbf{z}_j = \delta_{ij},$$
$$\mathbf{A}^{\mathsf{T}}\mathbf{w}_i = \sigma_i\mathbf{N}\mathbf{z}_i, \qquad \mathbf{w}_i^{\mathsf{T}}\mathbf{M}\mathbf{w}_j = \delta_{ij}.$$

In matrix form, this can also be cast as the generalized eigenvalue problem

$$\begin{bmatrix} & \mathbf{A} \\ \mathbf{A}^{\mathsf{T}} & \end{bmatrix}\begin{bmatrix} \mathbf{w} \\ \mathbf{z} \end{bmatrix} = \lambda\begin{bmatrix} \mathbf{M} & \\ & \mathbf{N} \end{bmatrix}\begin{bmatrix} \mathbf{w} \\ \mathbf{z} \end{bmatrix}.$$

The positive scalars σ_i are referred to as the elliptic singular values of $\mathbf{A}$, and the families $\{\mathbf{w}_i\}$ and $\{\mathbf{z}_i\}$ are its left and right elliptic singular vectors, respectively. The latter may equivalently be interpreted as the stationary points of the indefinite quadratic mapping

$$(\mathbf{w}, \mathbf{z}) \mapsto \mathbf{w}^{\mathsf{T}}\mathbf{A}\mathbf{z}$$

restricted to the unit sphere $\|\mathbf{w}\|_{\mathbf{M}} = 1$ and $\|\mathbf{z}\|_{\mathbf{N}} = 1$. In exact arithmetic, we have from (4.3) that $\bar{\mathbf{U}}_k^{\mathsf{T}}(\mathbf{M}^{-\frac{1}{2}}\mathbf{A}\mathbf{N}^{-\frac{1}{2}})\bar{\mathbf{V}}_k = \mathbf{B}_k$, where $\bar{\mathbf{u}}_j = \mathbf{M}^{\frac{1}{2}}\mathbf{u}_j$ and $\bar{\mathbf{v}}_j = \mathbf{N}^{\frac{1}{2}}\mathbf{v}_j$. Since the matrices $\bar{\mathbf{U}}_k$ and $\bar{\mathbf{V}}_k$ are orthogonal for all k, the singular values of $\mathbf{B}_{\min(m,n)}$ are the same as those of $\mathbf{M}^{-\frac{1}{2}}\mathbf{A}\mathbf{N}^{-\frac{1}{2}}$.

There are algebraically equivalent alternatives to Algorithm 4.2. For instance, if we use instead the change of variables $\mathbf{u}_i := \mathbf{M}^{\frac{1}{2}}\bar{\mathbf{u}}_i$ and $\mathbf{v}_i := \mathbf{N}^{-\frac{1}{2}}\bar{\mathbf{v}}_i$ and use the initial vector $\mathbf{M}^{\frac{1}{2}}\mathbf{b}$, we obtain the process described in Algorithm 4.3.

The process of Algorithm 4.3, which we could denote GKLB($\mathbf{M}^{-1}$, $\mathbf{N}$), generalizes the process referred to by Benbow (1999) as GKLB($\mathbf{M}^{-1}$). As both are mathematically equivalent, in the rest of this book, we concentrate on the process GKLB($\mathbf{M}$, $\mathbf{N}$) described by Algorithm 4.2. However, all methods examined in the next sections could be examined instead with Algorithm 4.3, and similar conclusions could be drawn.

Algorithm 4.3 Generalized Golub–Kahan Bidiagonalization, second variant

Require: $\mathbf{M}, \mathbf{A}, \mathbf{N}, \mathbf{b}$

1: $\beta_1 \mathbf{M}^{-1}\mathbf{u}_1 = \mathbf{b}$ with $\beta_1 > 0$ so that $\|\mathbf{u}_1\|_{\mathbf{M}^{-1}} = 1$
2: $\alpha_1 \mathbf{N}\mathbf{v}_1 = \mathbf{A}^T\mathbf{M}^{-1}\mathbf{u}_1$ with $\alpha_1 > 0$ so that $\|\mathbf{v}_1\|_{\mathbf{N}} = 1$
3: **for** $k = 1, 2, \dots$ **do**
4: $\quad \beta_{k+1}\mathbf{u}_{k+1} = \mathbf{A}\mathbf{v}_k - \alpha_k \mathbf{u}_k$ with $\beta_{k+1} > 0$ so that $\|\mathbf{u}_{k+1}\|_{\mathbf{M}^{-1}} = 1$
5: $\quad \alpha_{k+1}\mathbf{N}\mathbf{v}_{k+1} = \mathbf{A}^T\mathbf{M}^{-1}\mathbf{u}_{k+1} - \beta_{k+1}\mathbf{N}\mathbf{v}_k$ with $\alpha_{k+1} > 0$ so that $\|\mathbf{v}_{k+1}\|_{\mathbf{N}} = 1$.
6: **end for**

4.3 ▪ Relation between the Lanczos and Golub–Kahan Processes

The Lanczos and Golub–Kahan processes are intimately related, and several authors, including van der Vorst (2009) and Watkins (2007), clarify the connection between them. Although we are in possession of all the elements to do it here, we defer the description until Chapter 6, as it will allow us to deduce relations between methods based on those processes at the same time.

Chapter 5

Iterative Methods Based on Reduced Equations

This chapter presents a family of four methods based on the normal and Schur-complement equations. Three methods are generalizations of known methods in appropriate metrics: LSQR, CRAIG, and LSMR. The last one is new and may be viewed as an alternative to LSMR when $m < n$. It also serves as an essential tool to explain the behavior of MINRES on (1.5) in Section 6.4. For each method, we give implementation details in order to be complete and to provide a self-contained reference. The implementation details of LSQR and LSMR are our interpretation of the descriptions in Paige and Saunders (1982) and Fong and Saunders (2011). The implementation details of CRAIG were pieced together from various hints scattered across the literature, and those of CRAIG-MR are new, although they essentially mirror LSMR.

5.1 ▪ Generalized LSQR

The generalized LSQR method seeks a solution to the normal equations (2.6). At the end of the kth iteration of Algorithm 4.2, we may seek an approximation $\mathbf{y}_k$ to the solution $\mathbf{y}$ of (2.6) in the kth Krylov subspace spanned by $\{v_1, \dots, v_k\}$, i.e.,

$$\mathbf{y} \approx \mathbf{y}_k := \mathbf{V}_k \bar{\mathbf{y}}_k$$

for some $\bar{\mathbf{y}}_k \in \mathbb{R}^k$. Using (4.3), we have $\mathbf{A}\mathbf{y}_k = \mathbf{A}\mathbf{V}_k\bar{\mathbf{y}}_k = \mathbf{M}\mathbf{U}_{k+1}\mathbf{E}_k\bar{\mathbf{y}}_k$ in exact arithmetic so that

$$\mathbf{A}^\mathsf{T}\mathbf{M}^{-1}\mathbf{A}\mathbf{y}_k = (\mathbf{N}\mathbf{V}_k\mathbf{E}_k^\mathsf{T} + \alpha_{k+1}\mathbf{N}\mathbf{v}_{k+1}\mathbf{e}_{k+1}^\mathsf{T})\mathbf{E}_k\bar{\mathbf{y}}_k$$

and, using the definition of $\mathbf{y}_k$, $\mathbf{N}\mathbf{y}_k = \mathbf{N}\mathbf{V}_k\bar{\mathbf{y}}_k$. Similarly, the initialization of Algorithm 4.2 guarantees that $\mathbf{b} = \mathbf{M}\mathbf{U}_{k+1}(\beta_1\mathbf{e}_1)$ holds to machine precision at each iteration. This implies that

$$\mathbf{A}^\mathsf{T}\mathbf{M}^{-1}\mathbf{b} = (\mathbf{N}\mathbf{V}_k\mathbf{E}_k^\mathsf{T} + \alpha_{k+1}\mathbf{N}\mathbf{v}_{k+1}\mathbf{e}_{k+1}^\mathsf{T})(\beta_1\mathbf{e}_1).$$

Thus, (2.6) may be equivalently written as

$$((\mathbf{N}\mathbf{V}_k\mathbf{E}_k^\mathsf{T} + \alpha_{k+1}\mathbf{N}\mathbf{v}_{k+1}\mathbf{e}_{k+1}^\mathsf{T})\mathbf{E}_k + \mathbf{N}\mathbf{V}_k)\bar{\mathbf{y}}_k = (\mathbf{N}\mathbf{V}_k\mathbf{E}_k^\mathsf{T} + \alpha_{k+1}\mathbf{N}\mathbf{v}_{k+1}\mathbf{e}_{k+1}^\mathsf{T})(\beta_1\mathbf{e}_1).$$

Premultiplying the previous equality with $\mathbf{V}_k^\mathsf{T}$ and using the fact that the $\mathbf{v}_k$'s are $\mathbf{N}$-orthonormal in exact arithmetic, we obtain

$$(\mathbf{E}_k^\mathsf{T}\mathbf{E}_k + \mathbf{I}_k)\bar{\mathbf{y}}_k = \mathbf{E}_k^\mathsf{T}(\beta_1\mathbf{e}_1), \tag{5.1}$$

which are the optimality conditions of the linear least-squares problem

$$\underset{\bar{y}\in\mathbb{R}^k}{\text{minimize}}\ \frac{1}{2}\left\|\begin{bmatrix}E_k\\ I_k\end{bmatrix}\bar{y}-\begin{bmatrix}\beta_1 e_1\\ 0\end{bmatrix}\right\|_2^2. \tag{5.2}$$

The latter is exactly the kth regularized least-squares subproblem solved by LSQR with regularization parameter $\lambda = 1$. Equivalently, $\bar{y}_k$ solves the SQD subsystem

$$\begin{bmatrix}I_{k+1} & E_k\\ E_k^T & -I_k\end{bmatrix}\begin{bmatrix}\bar{x}_k\\ \bar{y}_k\end{bmatrix}=\begin{bmatrix}\beta_1 e_1\\ 0\end{bmatrix} \tag{5.3}$$

for some $\bar{x}_k$. As in Benbow (1999), this means that all that need be changed in LSQR to solve (1.5) with $f = b$ and $g = 0$ is Algorithm 4.1, which should be replaced with Algorithm 4.2.

The ℓ_2-norm residual satisifies

$$\|\bar{x}_k\|_2 = \|M^{\frac{1}{2}}U_{k+1}\bar{x}_k\|_2 = \|M^{-\frac{1}{2}}(Mx_k)\|_2 = \|Mx_k\|_{M^{-1}} = \|x_k\|_M,$$

where we used the fact that $M^{\frac{1}{2}}U_{k+1}$ is an orthonormal matrix. Thus, $\|Mx_k\|_{M^{-1}}$ may be updated recursively by updating $\|\bar{x}_k\|_2$ as in the original LSQR, and the sequence $\{\|Mx_k\|_{M^{-1}}\}$ is nonincreasing.

It is convenient to solve (5.2) by computing a $(2k+1)\times(2k+1)$ orthogonal matrix Q_k as a product of Givens reflections and a $(2k+1)\times k$ upper bidiagonal matrix R_k such that

$$\widetilde{E}_k := \begin{bmatrix}E_k\\ I_k\end{bmatrix} = Q_k R_k. \tag{5.4}$$

We give the details of the construction of Q_k in the next section.

The following result is algebraic and generalizes Saunders (1995, Result 8). It is based on the observation that the matrix of (2.6) may also be written as

$$N^{\frac{1}{2}}\left((M^{-\frac{1}{2}}AN^{-\frac{1}{2}})^T(M^{-\frac{1}{2}}AN^{-\frac{1}{2}})+I_n\right)N^{\frac{1}{2}}.$$

Theorem 5.1 (Relation between LSQR and CG). *The generalized* LSQR *iterates on* (2.4) *are the same as those generated by the standard conjugate-gradient method on the positive-definite system* (2.6) *with preconditioner* N.

Proof. Proceeding as above, using (4.3) and postmultiplying the matrix of (2.6) by V_k, we have

$$\begin{aligned}(A^TM^{-1}A+N)V_k &= N(V_kE_k^TE_k + \alpha_{k+1}v_{k+1}e_{k+1}^TE_k + V_k)\\ &= N(V_k(E_k^TE_k+I_k)+\alpha_{k+1}\beta_{k+1}v_{k+1}e_k^T),\end{aligned}$$

where we used the fact that $e_{k+1}^TE_k = \beta_{k+1}e_k^T$. The matrix $T_k := E_k^TE_k + I_k$ is tridiagonal, symmetric, and positive definite, and its ith off-diagonal element is $\alpha_i\beta_i$. Therefore, if we compare the above with (4.2), we see that it represents a Lanczos process applied to the matrix of (2.6) in a metric defined by N. Moreover, by definition of Q_k and R_k,

$$E_k^TE_k + I_k = \begin{bmatrix}E_k^T & I_k\end{bmatrix}Q_kQ_k^T\begin{bmatrix}E_k\\ I_k\end{bmatrix} = R_k^TR_k.$$

Therefore, $\mathbf{R}_k^\mathrm{T}$ is the Cholesky factor of $\mathbf{T}_k$, updated at each iteration, and the generalized LSQR method is equivalent to the method of conjugate gradients applied to (2.6) with preconditioner $\mathbf{N}$. □

5.2 ▪ Generalized LSQR: Implementation Details

In this section, we give update formulae to perform the factorization (5.4) iteratively. Substituting the identity

$$\mathbf{R}_k^\mathrm{T}\mathbf{R}_k\bar{\mathbf{y}}_k = \begin{bmatrix}\mathbf{E}_k^\mathrm{T} & \mathbf{I}_k\end{bmatrix}\begin{bmatrix}\mathbf{E}_k \\ \mathbf{I}_k\end{bmatrix}\bar{\mathbf{y}}_k = (\mathbf{E}_k^\mathrm{T}\mathbf{E}_k + \mathbf{I}_k)\bar{\mathbf{y}}_k$$

into (5.1), we obtain

$$\mathbf{R}_k^\mathrm{T}\mathbf{R}_k\bar{\mathbf{y}}_k = \mathbf{E}_k^\mathrm{T}\beta_1\mathbf{e}_1 = \begin{bmatrix}\mathbf{E}_k^\mathrm{T} & \mathbf{I}_k\end{bmatrix}\begin{bmatrix}\beta_1\mathbf{e}_1 \\ 0\end{bmatrix} = \mathbf{R}_k^\mathrm{T}\mathbf{Q}_k^\mathrm{T}\begin{bmatrix}\beta_1\mathbf{e}_1 \\ 0\end{bmatrix} = \alpha_1\beta_1\mathbf{e}_1. \tag{5.5}$$

The $(2j+1,j)$th element of $\widetilde{\mathbf{E}}_k$ (equal to 1) may be zeroed out by applying a Givens reflection acting on rows j and $2j$, denoted $\mathbf{Q}_{j,2j+1}^\mathrm{T}$, the last index indicating the row of the element being zeroed out. This reflection does not create any new nonzero in the other columns of $\widetilde{\mathbf{E}}_k$. Its effect may be represented schematically as (ignoring irrelevant rows and columns)

$$\begin{matrix} j \\ 2j+1 \end{matrix}\overset{\begin{matrix} j & \quad 2j+1 \end{matrix}}{\begin{bmatrix} c_j & s_j \\ s_j & -c_j \end{bmatrix}}\overset{\begin{matrix} j & \quad j+1 \end{matrix}}{\begin{bmatrix} \hat{\alpha}_j & 0 \\ 1 & 0 \end{bmatrix}} = \overset{\begin{matrix} j & \quad j+1 \end{matrix}}{\begin{bmatrix} \hat{\rho}_j & 0 \\ 0 & 0 \end{bmatrix}},$$

where $\hat{\rho}_j := \sqrt{\hat{\alpha}_j^2+1}$, $c_j := \hat{\alpha}_j/\hat{\rho}_j$, $s_j = 1/\hat{\rho}_j$, and, initially, $\hat{\alpha}_1 = \alpha_1$. Next, the β_{j+1} in position $(j+1,j)$ may be zeroed out by a second Givens reflection acting on rows j and $j+1$, denoted $\mathbf{Q}_{j,j+1}^\mathrm{T}$. This reflection creates a new nonzero element in position $(j,j+1)$ as the following schema illustrates:

$$\begin{matrix} j \\ j+1 \end{matrix}\overset{\begin{matrix} j & \quad j+1 \end{matrix}}{\begin{bmatrix} \bar{c}_j & \bar{s}_j \\ \bar{s}_j & -\bar{c}_j \end{bmatrix}}\overset{\begin{matrix} j & \quad j+1 \end{matrix}}{\begin{bmatrix} \hat{\rho}_j & 0 \\ \beta_{j+1} & \alpha_{j+1} \end{bmatrix}} = \overset{\begin{matrix} j & \quad j+1 \end{matrix}}{\begin{bmatrix} \rho_j & \theta_{j+1} \\ 0 & \hat{\alpha}_{j+1} \end{bmatrix}},$$

where $\rho_j := \sqrt{\hat{\rho}_j^2+\beta_{j+1}^2}$, $\bar{c}_j := \hat{\rho}_j/\rho_j$, $\bar{s}_j := \beta_{j+1}/\rho_j$, $\theta_{j+1} := \bar{s}_j\alpha_{j+1}$, and $\hat{\alpha}_{j+1} := -\bar{c}_j\alpha_{j+1}$. It is now easy to see that the overall orthogonal matrix $\mathbf{Q}_k$ is given by

$$\mathbf{Q}_k = (\mathbf{Q}_{1,4}\mathbf{Q}_{1,2})(\mathbf{Q}_{2,6}\mathbf{Q}_{2,3})\cdots(\mathbf{Q}_{k,2k}\mathbf{Q}_{k,k+1}).$$

Because $\mathbf{E}_k$ is $(k+1)\times k$, the result of the first k Givens reflections may be described as

$$\mathbf{Q}_k^\mathrm{T}\begin{bmatrix}\mathbf{E}_k & \beta_1\mathbf{e}_1 \\ \mathbf{I}_k & 0\end{bmatrix} = \mathbf{Q}_k^\mathrm{T}\begin{bmatrix}\mathbf{B}_k & \beta_1\mathbf{e}_1 \\ \beta_{k+1}\mathbf{e}_k^\mathrm{T} & 0 \\ \mathbf{I}_k & 0\end{bmatrix} = \begin{bmatrix}\mathbf{R}_k & \mathbf{z}_k \\ 0 & \bar{\zeta}_{k+1} \\ 0 & \mathbf{w}_k\end{bmatrix},$$

where $\mathbf{z}_k = (\zeta_1,\ldots,\zeta_k)$, $\mathbf{w}_k = (\omega_1,\ldots,\omega_k)$, and $\bar{\zeta}_{k+1}$ will be updated into ζ_{k+1} with the next Givens reflection.

The update of the right-hand side may be visualized as

$$\begin{array}{c} j \\ j+1 \\ 2j+1 \end{array}\begin{bmatrix} \bar{c}_j & \bar{s}_j & \\ \bar{s}_j & -\bar{c}_j & \\ & & 1 \end{bmatrix}\begin{bmatrix} c_j & & s_j \\ & 1 & \\ s_j & & -c_j \end{bmatrix}\begin{bmatrix} \bar{\zeta}_j \\ 0 \\ 0 \end{bmatrix} = \begin{bmatrix} \zeta_j \\ \bar{\zeta}_{j+1} \\ \omega_j \end{bmatrix},$$

where $\zeta_j := \bar{c}_j c_j \bar{\zeta}_j$, $\bar{\zeta}_{j+1} := \bar{s}_j c_j \bar{\zeta}_j$, $\omega_j := s_j \bar{\zeta}_j$ and where we initialize $\bar{\zeta}_1 := \beta_1$. The value $\bar{\zeta}_{j+1}$ will be replaced by ζ_{j+1} by the next Givens reflection.

A consequence of the reflection above is that the subproblem (5.2) may equivalently be rewritten as

$$\underset{\bar{y}\in\mathbb{R}^k}{\text{minimize}}\ \frac{1}{2}\left\|\begin{bmatrix} \mathbf{R}_k \\ 0 \\ 0 \end{bmatrix}\bar{\mathbf{y}} - \begin{bmatrix} \mathbf{z}_k \\ \bar{\zeta}_{k+1} \\ \mathbf{w}_k \end{bmatrix}\right\|_2^2,$$

whose solution is readily identified, $\bar{\mathbf{y}}_k := \mathbf{R}_k^{-1}\mathbf{z}_k$, and whose residual is the norm of $(\bar{\zeta}_{k+1}, \mathbf{w}_k)$.

Since $\bar{\mathbf{y}}_k$ is the solution of an upper triangular system, all of its components likely change at each iteration. Fortunately, it is possible to update $\mathbf{y}_k$ directly without requiring $\bar{\mathbf{y}}_k$. Following Paige and Saunders (1982), let $\mathbf{d}_j$ be the jth column of

$$\mathbf{D}_k := \mathbf{V}_k \mathbf{R}_k^{-1}. \tag{5.6}$$

Upon rearranging, we have $\mathbf{R}_k^\mathsf{T}\mathbf{D}_k^\mathsf{T} = \mathbf{V}_k^\mathsf{T}$ so that we find the rows of $\mathbf{D}_k^\mathsf{T}$, i.e., the vectors $\mathbf{d}_j$, recursively:

$$\mathbf{d}_1 = \frac{1}{\rho_1}\mathbf{v}_1, \qquad \mathbf{d}_{j+1} = \frac{1}{\rho_{j+1}}(\mathbf{v}_{j+1} - \theta_{j+1}\mathbf{d}_j) \quad (j = 1,\ldots,k-1). \tag{5.7}$$

Consequently,

$$\mathbf{y}_k = \mathbf{V}_k\bar{\mathbf{y}}_k = \mathbf{V}_k\mathbf{R}_k^{-1}\mathbf{z}_k = \mathbf{D}_k\mathbf{z}_k = \mathbf{y}_{k-1} + \zeta_k\mathbf{d}_k. \tag{5.8}$$

The following result shows that $\mathbf{D}_k$ is a partial factor of $\mathbf{A}^\mathsf{T}\mathbf{M}^{-1}\mathbf{A}+\mathbf{N}$ and that the latter matrix defines the appropriate norm to measure direct errors.

Theorem 5.2 (Solution and Error Norm in LSQR). *Let the vectors $\mathbf{d}_k$ be updated according to (5.7). Then, for $k = 1,\ldots,m$, we have*

$$\mathbf{D}_k^\mathsf{T}(\mathbf{A}^\mathsf{T}\mathbf{M}^{-1}\mathbf{A}+\mathbf{N})\mathbf{D}_k = \mathbf{I}_k. \tag{5.9}$$

In particular,

$$\mathbf{y}_k = \sum_{j=1}^{k}\zeta_j\mathbf{d}_j, \tag{5.10a}$$

$$\|\mathbf{y}\|^2_{\mathbf{A}^\mathsf{T}\mathbf{M}^{-1}\mathbf{A}+\mathbf{N}} = \sum_{j=1}^{m}\zeta_j^2, \tag{5.10b}$$

$$\|\mathbf{y}-\mathbf{y}_k\|^2_{\mathbf{A}^\mathsf{T}\mathbf{M}^{-1}\mathbf{A}+\mathbf{N}} = \sum_{j=k+1}^{m}\zeta_j^2, \tag{5.10c}$$

where $\mathbf{y}$ *is the solution of* (2.4).

Proof. We have from (4.3), (5.4), and (5.6) that

$$\begin{aligned}\mathbf{D}_k^\mathsf{T}(\mathbf{A}^\mathsf{T}\mathbf{M}^{-1}\mathbf{A}+\mathbf{N})\mathbf{D}_k &= \mathbf{R}_k^{-\mathsf{T}}\mathbf{V}_k^\mathsf{T}(\mathbf{A}^\mathsf{T}\mathbf{M}^{-1}\mathbf{A}+\mathbf{N})\mathbf{V}_k\mathbf{R}_k^{-1}\\ &= \mathbf{R}_k^{-\mathsf{T}}(\mathbf{V}_k^\mathsf{T}\mathbf{A}^\mathsf{T}\mathbf{M}^{-1}\mathbf{A}\mathbf{V}_k+\mathbf{I}_k)\mathbf{R}_k^{-1}\\ &= \mathbf{R}_k^{-\mathsf{T}}(\mathbf{E}_k^\mathsf{T}\mathbf{U}_{k+1}^\mathsf{T}\mathbf{M}\mathbf{U}_{k+1}\mathbf{E}_k+\mathbf{I}_k)\mathbf{R}_k^{-1}\\ &= \mathbf{R}_k^{-\mathsf{T}}\begin{bmatrix}\mathbf{E}_k^\mathsf{T} & \mathbf{I}_k\end{bmatrix}\begin{bmatrix}\mathbf{E}_k\\ \mathbf{I}_k\end{bmatrix}\mathbf{R}_k^{-1}\\ &= \mathbf{Q}_k^\mathsf{T}\mathbf{Q}_k = \mathbf{I}_k,\end{aligned}$$

which establishes (5.9). Formulae (5.10a), (5.10b), and (5.10c) follow easily from (5.9) and (5.8). □

Truncating the sum in (5.10b) and (5.10c) yields lower bounds on $\|\mathbf{y}\|^2_{\mathbf{A}^\mathsf{T}\mathbf{M}^{-1}\mathbf{A}+\mathbf{N}}$ and $\|\mathbf{y}-\mathbf{y}_k\|^2_{\mathbf{A}^\mathsf{T}\mathbf{M}^{-1}\mathbf{A}+\mathbf{N}}$. In particular, given an integer d sufficiently large and an iteration $k \ge d$, the sum over the most recent d iterations

$$\sum_{j=k-d+1}^{k} \zeta_j^2 \le \|\mathbf{y}-\mathbf{y}_{k-d+1}\|^2_{\mathbf{A}^\mathsf{T}\mathbf{M}^{-1}\mathbf{A}+\mathbf{N}} \tag{5.11}$$

can be a good approximation of the direct error between the exact solution $\mathbf{y}$ and the iterate $\mathbf{y}_{k-d+1}$ in the energy norm defined by $\mathbf{A}^\mathsf{T}\mathbf{M}^{-1}\mathbf{A}+\mathbf{N}$. Analogously to the approach used by Arioli (2013), Hestenes and Stiefel (1952), Golub and Meurant (1997, 2010), and others, this suggests a stopping criterion where we stop the iterations when the partial sum above falls below a tolerance τ^2 times $\sum_{j=1}^{k}\zeta_j^2$.

The update of $\mathbf{d}_{k+1}$ appears to require knowledge of ρ_{k+1}, which is not available during the kth iteration. As in Paige and Saunders (1982), it is possible to circumvent this by defining $\mathbf{h}_k := \rho_k\mathbf{d}_k$. We then initialize $\mathbf{h}_1 := \mathbf{v}_1$ and update according to $\mathbf{h}_{k+1} = \mathbf{v}_{k+1} - \theta_{k+1}/\rho_k\,\mathbf{h}_k$. The update of $\mathbf{y}_k$ becomes $\mathbf{y}_k = \mathbf{y}_{k-1} + \zeta_k/\rho_k\,\mathbf{h}_k$. The main steps of G-LSQR are summarized as Algorithm 5.1.

5.3 ▪ Generalized CRAIG

The generalized CRAIG method seeks a solution to the least-norm problem (2.8), which, we reiterate, is perfectly equivalent to the least-squares problem (2.4). After k steps of Algorithm 4.2, we seek approximations

$$\mathbf{x} \approx \mathbf{x}_k := \mathbf{U}_k\bar{\mathbf{x}}_k \qquad \text{and} \qquad \mathbf{y} \approx \mathbf{y}_k := \mathbf{V}_k\bar{\mathbf{y}}_k$$

for some $\bar{\mathbf{x}}_k \in \mathbb{R}^k$ and $\bar{\mathbf{y}}_k \in \mathbb{R}^k$. In doing so, we have $\|\mathbf{y}\|^2_\mathbf{N} \approx \|\mathbf{y}_k\|^2_\mathbf{N} = \|\bar{\mathbf{y}}_k\|^2$ and $\|\mathbf{x}\|^2_\mathbf{M} \approx \|\mathbf{x}_k\|^2_\mathbf{M} = \|\bar{\mathbf{x}}_k\|^2$. Moreover,

$$\begin{aligned}\mathbf{M}\mathbf{x}+\mathbf{A}\mathbf{y} &\approx \mathbf{M}\mathbf{x}_k+\mathbf{A}\mathbf{y}_k\\ &= \mathbf{M}\mathbf{U}_k\bar{\mathbf{x}}_k+\mathbf{A}\mathbf{V}_k\bar{\mathbf{y}}_k\\ &= \mathbf{M}\mathbf{U}_k\bar{\mathbf{x}}_k+\mathbf{M}\mathbf{U}_k\mathbf{B}_k\bar{\mathbf{y}}_k+\beta_{k+1}\mathbf{M}\mathbf{u}_{k+1}\mathbf{e}_k^\mathsf{T}\bar{\mathbf{y}}_k.\end{aligned}$$

Algorithm 5.1 Generalized LSQR

Require: $\mathbf{M}, \mathbf{A}, \mathbf{N}, \mathbf{b}, d, \tau, k_{max}$

$\beta_1 \mathbf{M}\mathbf{u}_1 = \mathbf{b}, \quad \alpha_1 \mathbf{N}\mathbf{v}_1 = \mathbf{A}^\mathsf{T}\mathbf{u}_1$ *// Initialize bidiagonalization*

$\mathbf{h}_1 = \mathbf{v}_1, \quad \bar{\zeta}_1 = \beta_1, \quad \hat{\alpha}_1 = \alpha_1, \quad \mathbf{y}_0 = 0$

$k = 1, \quad \Delta = 0,$ converged = **false**

while not converged **and** $k < k_{max}$ **do**

  // Continue bidiagonalization

  $\beta_{k+1}\mathbf{M}\mathbf{u}_{k+1} = \mathbf{A}\mathbf{v}_k - \alpha_k \mathbf{M}\mathbf{u}_k, \quad \alpha_{k+1}\mathbf{N}\mathbf{v}_{k+1} = \mathbf{A}^\mathsf{T}\mathbf{u}_{k+1} - \beta_{k+1}\mathbf{N}\mathbf{v}_k$

  $\hat{\rho}_k = (1 + \hat{\alpha}_k^2)^{\frac{1}{2}}, \quad c_k = \hat{\alpha}_k / \hat{\rho}_k, \quad s_k = 1/\hat{\rho}_k$ *// Reflection of type II*

  $\rho_k = (\hat{\rho}_k^2 + \beta_{k+1}^2)^{\frac{1}{2}}, \quad \bar{c}_k = \hat{\rho}_k / \rho_k, \quad \bar{s}_k = \beta_{k+1}/\rho_k$ *// Reflection of type I*

  $\theta_{k+1} = \bar{s}_k \alpha_{k+1}, \quad \hat{\alpha}_{k+1} = -\bar{c}_k \alpha_{k+1}$

  $\zeta_k = \bar{c}_k c_k \bar{\zeta}_k, \quad \bar{\zeta}_{k+1} = \bar{s}_k c_k \bar{\zeta}_k, \quad \omega_k = s_k \bar{\zeta}_k$ *// Update solution and residual*

  $\mathbf{y}_k = \mathbf{y}_{k-1} + \zeta_k / \rho_k \, \mathbf{h}_k$ *// Update*

  $\mathbf{h}_{k+1} = \mathbf{v}_{k+1} - \theta_{k+1}/\rho_k \, \mathbf{h}_k,$

  $\Delta = \Delta + \zeta_k^2$

  if $k \ge d$ **then**

    converged $= \left(\sum_{j=k-d+1}^{k} \zeta_j^2 < \tau^2 \Delta \right)$ *// Test convergence*

  end if

  $k \leftarrow k + 1$

end while

$\mathbf{y} = \mathbf{y}_k$

$\mathbf{x} = \mathbf{M}^{-1}(\mathbf{b} - \mathbf{A}\mathbf{y})$

return $(\mathbf{x}, \mathbf{y})$

After multiplication by $\mathbf{U}_k^\mathsf{T}$, the right-hand side becomes $\bar{\mathbf{x}}_k + \mathbf{B}_k \bar{\mathbf{y}}_k$. Therefore, $(\bar{\mathbf{x}}_k, \bar{\mathbf{y}}_k)$ solves

$$\underset{\bar{\mathbf{x}},\bar{\mathbf{y}}}{\text{minimize}} \ \tfrac{1}{2}(\|\bar{\mathbf{x}}\|^2 + \|\bar{\mathbf{y}}\|^2) \quad \text{subject to } \bar{\mathbf{x}} + \mathbf{B}_k \bar{\mathbf{y}} = \beta_1 \mathbf{e}_1, \tag{5.12}$$

where we again used the fact that $\mathbf{b} = \mathbf{M}\mathbf{U}_k(\beta_1 \mathbf{e}_1)$ for all k. Since the constraints of the latter problem always form a compatible system, Craig's method may be applied. As in the previous section, the kth subproblem solved by this generalized CRAIG method is identical to that solved by the *extended* CRAIG *algorithm* of Saunders (1995) with the regularization parameter set to one. The only part of the implementation that need be modified is the bidiagonalization step.

By contrast with Section 5.1, the solution of (5.12) solves the SQD subsystem

$$\begin{bmatrix} \mathbf{I}_k & \mathbf{B}_k \\ \mathbf{B}_k^\mathsf{T} & -\mathbf{I}_k \end{bmatrix} \begin{bmatrix} \bar{\mathbf{x}}_k \\ \bar{\mathbf{y}}_k \end{bmatrix} = \begin{bmatrix} \beta_1 \mathbf{e}_1 \\ 0 \end{bmatrix}, \tag{5.13}$$

which represents the optimality conditions of the regularized least-squares problem

$$\underset{\bar{\mathbf{y}} \in \mathbb{R}^k}{\text{minimize}} \ \frac{1}{2} \left\| \begin{bmatrix} \mathbf{B}_k \\ \mathbf{I}_k \end{bmatrix} \bar{\mathbf{y}} - \begin{bmatrix} \beta_1 \mathbf{e}_1 \\ 0 \end{bmatrix} \right\|_2^2 .$$

The similarity between the latter and the least-squares problem solved at iteration k of LSQR makes it possible to transfer from the CRAIG point to the LSQR point (Saunders, 1995).

Following Saunders (1995) and Paige (1974), (5.12) may be solved via the LQ factorization of the $k \times 2k$ matrix $\begin{bmatrix} \mathbf{B}_k & \mathbf{I}_k \end{bmatrix}$ by applying $2k-1$ Givens reflections that zero out the identity block. The construction of the reflections is explained below. Their effect is to produce an orthogonal $2k \times 2k$ matrix $\mathbf{Q}_k$ and a $k \times k$ lower bidiagonal matrix $\hat{\mathbf{B}}_k$ such that

$$\begin{bmatrix} \mathbf{B}_k & \mathbf{I}_k \end{bmatrix} \mathbf{Q}_k^{\mathsf{T}} = \begin{bmatrix} \hat{\mathbf{B}}_k & 0 \end{bmatrix}. \tag{5.14}$$

We denote

$$\hat{\mathbf{B}}_k := \begin{bmatrix} \hat{\alpha}_1 & & & \\ \hat{\beta}_2 & \hat{\alpha}_2 & & \\ & \ddots & \ddots & \\ & & \hat{\beta}_k & \hat{\alpha}_k \end{bmatrix}. \tag{5.15}$$

Suppose $\mathbf{y}^*$ is the $\mathbf{y}$-component of the exact solution to (1.5) with $\mathbf{f} = \mathbf{b}$ and $\mathbf{g} = 0$ and write $\mathbf{y}^* = \mathbf{V}_m \bar{\mathbf{y}}$ for some vector $\bar{\mathbf{y}}$. The direct error is measured by

$$\|\bar{\mathbf{y}} - \bar{\mathbf{y}}_k\|_2 = \|\mathbf{N}^{\frac{1}{2}} \mathbf{V}_m (\bar{\mathbf{y}} - \mathbf{B}_k^{\mathsf{T}} \bar{\mathbf{z}}_k)\|_2 = \|\mathbf{V}_m (\bar{\mathbf{y}} - \mathbf{B}_k^{\mathsf{T}} \bar{\mathbf{z}}_k)\|_{\mathbf{N}} = \|\mathbf{y}^* - \mathbf{y}_k\|_{\mathbf{N}},$$

where we used the facts that $\bar{\mathbf{y}}_k = \mathbf{B}_k^{\mathsf{T}} \bar{\mathbf{z}}_k$, and $\mathbf{y}_k = \mathbf{V}_m \bar{\mathbf{y}}_k$ and the orthogonality of $\mathbf{N}^{\frac{1}{2}} \mathbf{V}_m$. Similarly, $\|\mathbf{x}^* - \mathbf{x}_k\|_{\mathbf{M}} = \|\bar{\mathbf{x}} - \bar{\mathbf{x}}_k\|_2$, where $\mathbf{x}^* = \mathbf{U}_m \bar{\mathbf{x}}$ is the $\mathbf{x}$-component of the exact solution to (1.5). The generalized CRAIG method thus generates a sequence $\{(\mathbf{x}_k, \mathbf{y}_k)\}$ such that $\|\mathbf{x}^* - \mathbf{x}_k\|_{\mathbf{M}}^2 + \|\mathbf{y}^* - \mathbf{y}_k\|_{\mathbf{N}}^2$ is nonincreasing.

By definition of the CRAIG method (Craig (1955); Saunders (1995, Result 9)) and the observation that

$$\mathbf{M} + \mathbf{A}\mathbf{N}^{-1}\mathbf{A}^{\mathsf{T}} = \mathbf{M}^{\frac{1}{2}} \left(\mathbf{I}_n + (\mathbf{M}^{-\frac{1}{2}} \mathbf{A} \mathbf{N}^{-\frac{1}{2}})(\mathbf{M}^{-\frac{1}{2}} \mathbf{A} \mathbf{N}^{-\frac{1}{2}})^{\mathsf{T}} \right) \mathbf{M}^{\frac{1}{2}},$$

we have the following result.

Theorem 5.3 (Relation between CRAIG and CG). *The generalized* CRAIG *iterates* $\mathbf{y}_k$ *are related to the iterates* $\mathbf{x}_k$ *of the conjugate-gradient method applied to*

$$(\mathbf{M} + \mathbf{A}\mathbf{N}^{-1}\mathbf{A}^{\mathsf{T}})\mathbf{x} = \mathbf{b} \tag{5.16}$$

with preconditioner $\mathbf{M}$ *according to* $\mathbf{y}_k = \mathbf{N}^{-1}\mathbf{A}^{\mathsf{T}}\mathbf{x}_k$.

Proof. Upon multiplying the second block equation of (5.13) by $\mathbf{B}_k$ and substituting the value of $\mathbf{B}_k \bar{\mathbf{y}}_k$ from the first block equation, we obtain

$$(\mathbf{B}_k \mathbf{B}_k^{\mathsf{T}} + \mathbf{I}_k) \bar{\mathbf{x}}_k = \beta_1 \mathbf{e}_1. \tag{5.17}$$

Our substitution, combined with (4.3), yields

$$\mathbf{y}_k = \mathbf{V}_k \bar{\mathbf{y}}_k = \mathbf{V}_k \mathbf{B}_k^{\mathsf{T}} \bar{\mathbf{x}}_k = \mathbf{N}^{-1} \mathbf{B}^{\mathsf{T}} \mathbf{U}_k \bar{\mathbf{x}}_k = \mathbf{N}^{-1} \mathbf{A}^{\mathsf{T}} \mathbf{x}_k.$$

Using the approximation $\mathbf{x} \approx \mathbf{x}_k := \mathbf{U}_k \bar{\mathbf{x}}_k$ in (5.16) and premultiplying with $\mathbf{U}_k^{\mathsf{T}}$, and using the $\mathbf{M}$-orthogonality of the $\mathbf{u}_k$, the $\mathbf{N}$-orthogonality of the $\mathbf{v}_k$, and (4.3), we obtain precisely (5.17).

System (5.17) may be written equivalently as

$$\begin{bmatrix} \mathbf{B}_k & \mathbf{I}_k \end{bmatrix} \begin{bmatrix} \mathbf{B}_k^\mathsf{T} \\ \mathbf{I}_k \end{bmatrix} \bar{\mathbf{x}}_k = \beta_1 \mathbf{e}_1.$$

Substituting the factorization (5.14) and using the orthogonality of $\mathbf{Q}_k$, we obtain

$$\hat{\mathbf{B}}_k \hat{\mathbf{B}}_k^\mathsf{T} \bar{\mathbf{x}}_k = \beta_1 \mathbf{e}_1. \tag{5.18}$$

In the latter system, the matrix $\hat{\mathbf{T}}_k := \hat{\mathbf{B}}_k \hat{\mathbf{B}}_k^\mathsf{T}$ is tridiagonal, symmetric, and positive definite, and its Cholesky factor is $\hat{\mathbf{B}}_k$. It is the tridiagonal matrix generated by a Lanczos process applied to (5.16). Indeed, consider (5.16), in which we substitute $\mathbf{x}$ by an approximation of the form $\mathbf{x}_k = \mathbf{U}_k \bar{\mathbf{x}}_k$. Using (4.3), we have

$$\begin{aligned} (\mathbf{M} + \mathbf{A}\mathbf{N}^{-1}\mathbf{A}^\mathsf{T})\mathbf{U}_k &= \mathbf{M}\big(\mathbf{U}_k(\mathbf{I}_k + \mathbf{B}_k \mathbf{B}_k^\mathsf{T}) + \beta_{k+1}\mathbf{u}_{k+1}\mathbf{e}_k^\mathsf{T}\mathbf{B}_k^\mathsf{T}\big) \\ &= \mathbf{M}\left(\mathbf{U}_k \begin{bmatrix} \mathbf{B}_k & \mathbf{I}_k \end{bmatrix} \begin{bmatrix} \mathbf{B}_k^\mathsf{T} \\ \mathbf{I}_k \end{bmatrix} + \alpha_k \beta_{k+1} \mathbf{u}_{k+1} \mathbf{e}_k^\mathsf{T}\right). \end{aligned} \tag{5.19}$$

It is easy to verify that the kth off-diagonal element of $\hat{\mathbf{T}}_k = \mathbf{B}_k \mathbf{B}_k^\mathsf{T} + \mathbf{I}_k$ is precisely equal to $\alpha_k \beta_{k+1}$. Comparing with (4.2), we conclude that (5.19) corresponds to a Lanczos process on the matrix of (5.16) with a metric defined by $\mathbf{M}$. This and the facts established above confirm that the generalized CRAIG method is equivalent to the conjugate-gradient method applied to (5.16) with preconditioner $\mathbf{M}$. □

The proof of Theorem 5.3, and (5.18) in particular, suggests a numerical procedure. Indeed, with

$$\bar{\mathbf{z}}_k := \hat{\mathbf{B}}_k^\mathsf{T} \bar{\mathbf{x}}_k, \tag{5.20}$$

the components of $\bar{\mathbf{z}}_k = (\zeta_1, \dots, \zeta_k)$ are given by the recursion

$$\zeta_1 := \beta_1 / \hat{\alpha}_1, \quad \zeta_{i+1} := -\hat{\beta}_{i+1} \zeta_i / \hat{\alpha}_i, \quad (i = 1, \dots, k-1). \tag{5.21}$$

Solving for $\mathbf{x}_k$ directly and bypassing $\bar{\mathbf{x}}_k$ may now be done as in Section 5.2. Define the columns $\mathbf{d}_i$ of $\mathbf{D}_k$ via $\hat{\mathbf{B}}_k \mathbf{D}_k^\mathsf{T} := \mathbf{U}_k^\mathsf{T}$. Then

$$\mathbf{x}_k = \mathbf{U}_k \bar{\mathbf{x}}_k = \mathbf{D}_k \hat{\mathbf{B}}_k^\mathsf{T} \bar{\mathbf{x}}_k = \mathbf{D}_k \bar{\mathbf{z}}_k.$$

The columns $\mathbf{d}_i$ are found recursively using

$$\mathbf{d}_1 := \mathbf{u}_1 / \hat{\alpha}_1, \quad \mathbf{d}_{i+1} := (\mathbf{u}_{i+1} - \hat{\beta}_{i+1} \mathbf{d}_i) / \hat{\alpha}_{i+1} \quad (i = 1, \dots, k-1). \tag{5.22}$$

This yields the update

$$\mathbf{x}_{k+1} = \mathbf{x}_k + \zeta_{k+1} \mathbf{d}_{k+1} \tag{5.23}$$

for $\mathbf{x}_{k+1}$. In the next section, we return to the expansion of each $\mathbf{x}_k$ as a linear combination of the columns of $\mathbf{D}_k$.

5.4 ▪ Generalized CRAIG: Implementation Details

Saunders (1995) describes an implementation of the extended CRAIG method in the variable $\mathbf{y}$. In this section, we describe an implementation in the variable $\mathbf{x}$ that resembles that of Arioli (2013).

The $(1, k+1)$ element of $\begin{bmatrix} \mathbf{B}_k & \mathbf{I}_k \end{bmatrix}$ (equal to 1) may be zeroed out by applying a Givens reflection acting on columns 1 and $k+1$. By convention, we denote this reflection $\mathbf{Q}_{1,k+1}$, the last index indicating the column of the element being zeroed out, and label it a reflection of type I. As the $(2,1)$ element of the constraint matrix is $\beta_2 > 0$, the reflection creates a new nonzero element in position $(2, k+1)$, which we denote γ_2. The newly created γ_2 may be zeroed out by a second reflection, of type II, acting on columns $k+1$ and $k+2$, i.e., $\mathbf{Q}_{k+2,k+1}$. Aside from zeroing out γ_2, the effect of $\mathbf{Q}_{k+2,k+1}$ is to change the value 1 in position $(k+2, k+2)$ to some other value, denoted δ_2, which can be zeroed out by a new reflection of type I. With the convention that $\delta_1 = 1$, a general reflection of type I, with the purpose of zeroing out a δ_k, may be represented schematically as (ignoring all irrelevant rows and columns)

$$\begin{array}{c} \\ k \\ k+1 \end{array}\!\!\begin{array}{c} \begin{matrix} k & \quad 2k \end{matrix} \\ \begin{bmatrix} \alpha_k & \delta_k \\ \beta_{k+1} & 0 \end{bmatrix} \end{array} \begin{array}{c} \begin{matrix} k & \quad 2k \end{matrix} \\ \begin{bmatrix} c_k & s_k \\ s_k & -c_k \end{bmatrix} \end{array} = \begin{bmatrix} \rho_k & 0 \\ c_k \beta_{k+1} & \gamma_{k+1} \end{bmatrix} := \begin{bmatrix} \hat{\alpha}_k & 0 \\ \hat{\beta}_{k+1} & \gamma_{k+1} \end{bmatrix},$$

where $\rho_k := \sqrt{\alpha_k^2 + \delta_k^2}$, $c_k := \alpha_k / \rho_k$, $s_k := \delta_k / \rho_k$, $\hat{\beta}_{k+1} := c_k \beta_{k+1}$, and $\gamma_{k+1} := s_k \beta_{k+1}$ and where labels to the left and above a matrix indicate row and column indices, respectively. It is easy to show by induction that $\hat{\alpha}_k > 0$ and therefore that the procedure (5.21) is well defined. Similarly, a general reflection of type II, with the purpose of zeroing out a γ_{k+1}, may be represented schematically as

$$\begin{array}{c} \\ k+1 \\ k+2 \end{array}\!\!\begin{array}{c} \begin{matrix} 2k+1 & \quad 2k+2 \end{matrix} \\ \begin{bmatrix} \gamma_{k+1} & 1 \\ 0 & 0 \end{bmatrix} \end{array} \begin{array}{c} \begin{matrix} 2k+1 & \quad 2k+2 \end{matrix} \\ \begin{bmatrix} \bar{c}_k & \bar{s}_k \\ \bar{s}_k & -\bar{c}_k \end{bmatrix} \end{array} = \begin{bmatrix} 0 & \delta_{k+1} \\ 0 & 0 \end{bmatrix},$$

where $\bar{c}_k := -1/\sqrt{\gamma_{k+1}^2 + 1}$, $\bar{s}_k := \gamma_{k+1}/\sqrt{\gamma_{k+1}^2 + 1}$, and $\delta_{k+1} := \bar{s}_k \gamma_{k+1} - \bar{c}_k = \sqrt{\gamma_{k+1}^2 + 1}$. The sequence of reflections required to perform the LQ factorization is given by

$$\mathbf{Q}_k^{\mathrm{T}} := (\mathbf{Q}_{1,k+1} \mathbf{Q}_{k+2,k+1})(\mathbf{Q}_{2,k+2} \mathbf{Q}_{k+3,k+2}) \cdots (\mathbf{Q}_{k-1,2k-1} \mathbf{Q}_{2k,2k-1}) \mathbf{Q}_{k,2k}.$$

Other reflections are possible; the one given coincides with that of Saunders (1995).

We can verify by induction that

$$\begin{aligned} \hat{\alpha}_1 &= \sqrt{\alpha_1^2 + 1}, \\ \hat{\alpha}_{k+1} &= \sqrt{\alpha_{k+1}^2 + \delta_{k+1}^2}, \quad k \geq 2, \\ \hat{\beta}_{k+1} &= \alpha_k \beta_{k+1} / \hat{\alpha}_k, \quad k \geq 2. \end{aligned} \tag{5.24}$$

At this point we have constructed $\hat{\mathbf{B}}_k$, and we may update the vectors $\mathbf{d}_k$ and the estimate $\mathbf{x}_k$ as in (5.22) and (5.23).

The factorization (5.14) implicitly fixes $\bar{\mathbf{y}}_k = \mathbf{B}_k^{\mathrm{T}} \bar{\mathbf{x}}_k$ in (5.13).

At each iteration, the residual of (1.5) at $(\mathbf{x}_k, \mathbf{y}_k)$,

$$\begin{bmatrix} \mathbf{r}_k \\ \mathbf{s}_k \end{bmatrix} := \begin{bmatrix} \mathbf{b} \\ 0 \end{bmatrix} - \begin{bmatrix} \mathbf{M} & \mathbf{A} \\ \mathbf{A}^\mathrm{T} & -\mathbf{N} \end{bmatrix} \begin{bmatrix} \mathbf{x}_k \\ \mathbf{y}_k \end{bmatrix}, \tag{5.25}$$

is given by

$$\begin{aligned}
\begin{bmatrix} \mathbf{r}_k \\ \mathbf{s}_k \end{bmatrix} &= \begin{bmatrix} \beta_1 \mathbf{M}\mathbf{u}_1 \\ 0 \end{bmatrix} - \begin{bmatrix} \mathbf{M} & \mathbf{A} \\ \mathbf{A}^\mathrm{T} & -\mathbf{N} \end{bmatrix} \begin{bmatrix} \mathbf{U}_k & \\ & \mathbf{V}_k \end{bmatrix} \begin{bmatrix} \bar{\mathbf{x}}_k \\ \bar{\mathbf{y}}_k \end{bmatrix} \\
&= \begin{bmatrix} \beta_1 \mathbf{M}\mathbf{u}_1 \\ 0 \end{bmatrix} - \begin{bmatrix} \mathbf{M} & \\ & \mathbf{N} \end{bmatrix} \begin{bmatrix} \mathbf{U}_k & \\ & \mathbf{V}_k \end{bmatrix} \begin{bmatrix} \mathbf{I}_k & \mathbf{B}_k \\ \mathbf{B}_k^T & -\mathbf{I}_k \end{bmatrix} \begin{bmatrix} \bar{\mathbf{x}}_k \\ \bar{\mathbf{y}}_k \end{bmatrix} - \begin{bmatrix} \beta_{k+1} \mathbf{M}\mathbf{u}_{k+1} \\ 0 \end{bmatrix} \mathbf{e}_{2k}^\mathrm{T} \begin{bmatrix} \bar{\mathbf{x}}_k \\ \bar{\mathbf{y}}_k \end{bmatrix} \\
&= \begin{bmatrix} -\beta_{k+1} \eta_k \mathbf{M}\mathbf{u}_{k+1} \\ 0 \end{bmatrix},
\end{aligned}$$

where we used (4.3) and (5.13) and where η_k is the kth component of $\bar{\mathbf{y}}_k$. The vector $\bar{\mathbf{y}}_k$ is not directly available, but η_k can nonetheless be monitored cheaply because, by definition,

$$\eta_k = \mathbf{e}_k^\mathrm{T} \begin{bmatrix} \bar{\mathbf{y}}_k \\ \bar{\mathbf{x}}_k \end{bmatrix} = \mathbf{e}_k^\mathrm{T} \mathbf{Q}_k^\mathrm{T} \begin{bmatrix} \bar{\mathbf{z}}_k \\ 0 \end{bmatrix} = \mathbf{e}_k^\mathrm{T} \mathbf{Q}_{k,2k} \begin{bmatrix} \bar{\mathbf{z}}_k \\ 0 \end{bmatrix} = c_k \zeta_k, \tag{5.26}$$

where ζ_k is the kth component of $\bar{\mathbf{z}}_k$, while c_k is the cosine defining $\mathbf{Q}_{k,2k}$. This establishes that

$$\|\mathbf{r}_k\|_{\mathbf{M}^{-1}} = |\beta_{k+1} c_k \zeta_k|. \tag{5.27}$$

The residual of (5.16) may be monitored similarly. As in (5.19), we have

$$\begin{aligned}
\mathbf{q}_k &:= \mathbf{b} - (\mathbf{M} + \mathbf{A}\mathbf{N}^{-1}\mathbf{A}^\mathrm{T}) \mathbf{U}_k \bar{\mathbf{x}}_k \\
&= \beta_1 \mathbf{M}\mathbf{u}_1 - \mathbf{M}\big(\mathbf{U}_k(\mathbf{I}_k + \mathbf{B}_k\mathbf{B}_k^\mathrm{T})\bar{\mathbf{x}}_k + \alpha_k \beta_{k+1} \mathbf{u}_{k+1} \mathbf{e}_k^\mathrm{T} \bar{\mathbf{x}}_k\big) \\
&= \alpha_k \beta_{k+1} \xi_k \mathbf{M}\mathbf{u}_{k+1},
\end{aligned} \tag{5.28}$$

where we used (5.17) and ξ_k is the last component of $\bar{\mathbf{x}}_k$. As before, $\bar{\mathbf{x}}_k$ is not directly accessible but, by definition,

$$\xi_k = \mathbf{e}_{2k}^\mathrm{T} \begin{bmatrix} \bar{\mathbf{y}}_k \\ \bar{\mathbf{x}}_k \end{bmatrix} = \mathbf{e}_{2k}^\mathrm{T} \mathbf{Q}_k^\mathrm{T} \begin{bmatrix} \bar{\mathbf{z}}_k \\ 0 \end{bmatrix} = \mathbf{e}_{2k}^\mathrm{T} \mathbf{Q}_{2k,2k-1} \mathbf{Q}_{k,2k} \begin{bmatrix} \bar{\mathbf{z}}_k \\ 0 \end{bmatrix} = s_k \zeta_k, \tag{5.29}$$

where we have used the facts that $\mathbf{e}_{2k}^\mathrm{T} \mathbf{Q}_{2k,2k-1} = \bar{s}_k \mathbf{e}_{2k-1}^\mathrm{T} - \bar{c}_k \mathbf{e}_{2k}^\mathrm{T}$, that $\mathbf{e}_{2k-1}^\mathrm{T} \mathbf{Q}_{k,2k} = 0$, and $\mathbf{e}_{2k}^\mathrm{T} \mathbf{Q}_{k,2k} = s_k \mathbf{e}_k^\mathrm{T} - c_k \mathbf{e}_{2k}^\mathrm{T}$. Since $\mathbf{q}_k \in \mathbb{M}^*$, we have, using the $\mathbf{M}$-orthogonality of $\mathbf{u}_{k+1}$,

$$\|\mathbf{q}_k\|_{\mathbf{M}^{-1}} = |\alpha_k \beta_{k+1} s_k \zeta_k|. \tag{5.30}$$

Note that from (5.21), (5.26), and (5.29), we also have

$$\xi_k^2 + \eta_k^2 = \zeta_k^2 \quad \text{for all } k. \tag{5.31}$$

The generalized CRAIG method is summarized as Algorithm 5.2.

As in the case of LSQR, Algorithm 5.2 implicitly constructs and updates a partial factorization of a matrix determining the energy norm.

Theorem 5.4 (Solution and Error Norm in CRAIG). *Let* $\hat{\mathbf{B}}_k$ *be defined by* (5.14) *and* $\mathbf{D}_k := \mathbf{U}_k\hat{\mathbf{B}}_k^{-\mathsf{T}}$. *For* $k = 1,\dots,n$, *we have*

$$\mathbf{D}_k^{\mathsf{T}}(\mathbf{A}\mathbf{N}^{-1}\mathbf{A}^{\mathsf{T}} + \mathbf{M})\mathbf{D}_k = \mathbf{I}_k. \tag{5.32}$$

In particular,

$$\mathbf{x}_k = \sum_{j=1}^{k} \zeta_j \mathbf{d}_j, \tag{5.33}$$

and we have the estimates

$$\|\mathbf{x}_k\|^2_{\mathbf{A}\mathbf{N}^{-1}\mathbf{A}^{\mathsf{T}}+\mathbf{M}} = \sum_{i=1}^{k} \zeta_i^2, \tag{5.34a}$$

$$\|\mathbf{x}^* - \mathbf{x}_k\|^2_{\mathbf{A}\mathbf{N}^{-1}\mathbf{A}^{\mathsf{T}}+\mathbf{M}} = \sum_{i=k+1}^{n} \zeta_i^2, \tag{5.34b}$$

$$\|\mathbf{x}_k\|^2_{\mathbf{M}} = \sum_{i=1}^{k} \xi_i^2, \tag{5.34c}$$

$$\|\mathbf{x}^* - \mathbf{x}_k\|^2_{\mathbf{M}} = \sum_{i=k+1}^{n} \xi_i^2, \tag{5.34d}$$

where ζ_i *and* ξ_i *are as in* (5.21) *and* (5.29).

In addition, the residuals $\mathbf{r}_k$ *and* $\mathbf{q}_k$ *defined in* (5.25) *and* (5.28) *satisfy*

$$\|\mathbf{r}_k\|_{\mathbf{M}^{-1}} \le \|\bar{\mathbf{A}}\|_2\,|\zeta_k| \quad \textit{and} \quad \|\mathbf{q}_k\|_{\mathbf{M}^{-1}} \le \|\bar{\mathbf{A}}\|_2^2\,|\zeta_k|. \tag{5.35}$$

Proof. We have from (5.14), the orthogonality of $\mathbf{Q}_k$, the $\mathbf{M}$-orthogonality of the $\mathbf{u}_k$, the $\mathbf{N}$-orthogonality of the $\mathbf{v}_k$, and (4.3) that

$$\begin{aligned}
\mathbf{D}_k^{\mathsf{T}}(\mathbf{A}\mathbf{N}^{-1}\mathbf{A}^{\mathsf{T}} + \mathbf{M})\mathbf{D}_k &= \hat{\mathbf{B}}_k^{-1}\mathbf{U}_k^{\mathsf{T}}(\mathbf{A}\mathbf{N}^{-1}\mathbf{A}^{\mathsf{T}} + \mathbf{M})\mathbf{U}_k\hat{\mathbf{B}}_k^{-\mathsf{T}} \\
&= \hat{\mathbf{B}}_k^{-1}(\mathbf{U}_k^{\mathsf{T}}\mathbf{A}\mathbf{N}^{-1}\mathbf{A}^{\mathsf{T}}\mathbf{U}_k + \mathbf{I}_k)\hat{\mathbf{B}}_k^{-\mathsf{T}} \\
&= \hat{\mathbf{B}}_k^{-1}(\mathbf{B}_k\mathbf{V}_k^{\mathsf{T}}\mathbf{N}\mathbf{V}_k\mathbf{B}_k^{\mathsf{T}} + \mathbf{I}_k)\hat{\mathbf{B}}_k^{-\mathsf{T}} \\
&= \hat{\mathbf{B}}_k^{-1}(\mathbf{B}_k\mathbf{B}_k^{\mathsf{T}} + \mathbf{I}_k)\hat{\mathbf{B}}_k^{-\mathsf{T}} \\
&= \hat{\mathbf{B}}_k^{-1}\begin{bmatrix}\mathbf{B}_k & \mathbf{I}_k\end{bmatrix}\begin{bmatrix}\mathbf{B}_k^{\mathsf{T}} \\ \mathbf{I}_k\end{bmatrix}\hat{\mathbf{B}}_k^{-\mathsf{T}} \\
&= \hat{\mathbf{B}}_k^{-1}\begin{bmatrix}\hat{\mathbf{B}}_k & 0\end{bmatrix}\mathbf{Q}_k\mathbf{Q}_k^{\mathsf{T}}\begin{bmatrix}\hat{\mathbf{B}}_k^{\mathsf{T}} \\ 0\end{bmatrix}\hat{\mathbf{B}}_k^{-\mathsf{T}} \\
&= \hat{\mathbf{B}}_k^{-1}\hat{\mathbf{B}}_k\hat{\mathbf{B}}_k^{\mathsf{T}}\hat{\mathbf{B}}_k^{-\mathsf{T}} \\
&= \mathbf{I}_k,
\end{aligned}$$

which establishes (5.32).

Using the update formula (5.23) for $\mathbf{x}_k$, we may write $\mathbf{x}_k = \mathbf{D}_k\bar{\mathbf{z}}_k$. Thus,

$$\|\mathbf{x}_k\|^2_{\mathbf{A}\mathbf{N}^{-1}\mathbf{A}^{\mathsf{T}}+\mathbf{M}} = \mathbf{x}_k^{\mathsf{T}}(\mathbf{A}\mathbf{N}^{-1}\mathbf{A}^{\mathsf{T}} + \mathbf{M})\mathbf{x}_k = \bar{\mathbf{z}}_k^{\mathsf{T}}\mathbf{D}_k^{\mathsf{T}}(\mathbf{A}\mathbf{N}^{-1}\mathbf{A}^{\mathsf{T}} + \mathbf{M})\mathbf{D}_k\bar{\mathbf{z}}_k = \|\bar{\mathbf{z}}_k\|_2^2.$$

Therefore,

$$\|\mathbf{x}_k\|^2_{\mathbf{AN}^{-1}\mathbf{A}^\mathsf{T}+\mathbf{M}} = \sum_{i=1}^{k} \zeta_i^2,$$

and this establishes (5.34a). Since

$$\|\mathbf{x}^*\|^2_{\mathbf{AN}^{-1}\mathbf{A}^\mathsf{T}+\mathbf{M}} = \|\mathbf{x}_n\|^2_{\mathbf{AN}^{-1}\mathbf{A}^\mathsf{T}+\mathbf{M}} = \sum_{i=1}^{n} \zeta_i^2,$$

we obtain (5.34b) as before.

Since $\mathbf{Mx}_k = \mathbf{MU}_k\bar{\mathbf{x}}_k$ and $\|\bar{\mathbf{x}}_k\|_2^2 = \sum_{i=1}^{k} \xi_i^2$ by definition of ξ_i, we obtain the second equality in (5.34c).

Using

$$\|\mathbf{x}^*\|^2_{\mathbf{M}} = \|\mathbf{x}_n\|^2_{\mathbf{M}} = \sum_{i=1}^{n} \xi_i^2,$$

we have the error estimate

$$\|\mathbf{x}^* - \mathbf{x}_k\|^2_{\mathbf{M}} = \sum_{i=k+1}^{n} \xi_i^2,$$

which is identical to (5.34d).

The proof of (5.35) follows directly from (5.25), (5.28), and Arioli (2013, Proposition 3.1). □

As in Section 5.2, (5.34b) suggests a stopping condition of the form

$$\sum_{i=k+1}^{k+d+1} \zeta_i^2 \le \tau^2 \sum_{i=1}^{k+d+1} \zeta_i^2$$

for a user-chosen integer $d \in \mathbb{N}_0$ and $\tau \in (0,1)$.

In Theorem 5.4, the quantity $\|\overline{\mathbf{A}}\|_2$ is the largest singular value of $\overline{\mathbf{A}}$, which coincides with the largest elliptic singular value of $\mathbf{A}$.

5.5 ▪ Generalized LSMR

LSMR (Fong and Saunders, 2011) consists in applying MINRES (Paige and Saunders, 1975) to the normal equations (2.6). Since the appropriate norm for measuring the residual of the normal equations is the $\mathbf{N}^{-1}$-norm, we premultiply (2.6) by $\mathbf{N}^{-\frac{1}{2}}$:

$$\mathbf{N}^{-\frac{1}{2}}(\mathbf{A}^\mathsf{T}\mathbf{M}^{-1}\mathbf{A}+\mathbf{N})\mathbf{y} = \mathbf{N}^{-\frac{1}{2}}\mathbf{A}^\mathsf{T}\mathbf{M}^{-1}\mathbf{b}. \tag{5.36}$$

By definition, MINRES then computes $\mathbf{y}$ so as to

$$\underset{\mathbf{y}\in\mathbb{R}^m}{\text{minimize}}\ \tfrac{1}{2}\|\mathbf{N}^{-\frac{1}{2}}(\mathbf{A}^\mathsf{T}\mathbf{M}^{-1}\mathbf{b} - (\mathbf{A}^\mathsf{T}\mathbf{M}^{-1}\mathbf{A}+\mathbf{N})\mathbf{y}))\|_2.$$

Seeking again an approximation $\mathbf{y} \approx \mathbf{y}_k := \mathbf{V}_k\bar{\mathbf{y}}_k$ and using (4.3), we have

$$\mathbf{M}^{-1}(\mathbf{b}-\mathbf{Ay}) = \mathbf{M}^{-1}(\mathbf{MU}_{k+1}(\beta_1\mathbf{e}_1) - \mathbf{AV}_k\bar{\mathbf{y}}_k) = \mathbf{U}_{k+1}(\beta_1\mathbf{e}_1 - \mathbf{E}_k\bar{\mathbf{y}}_k)$$

Algorithm 5.2 Generalized CRAIG

Require: $\mathbf{M}, \mathbf{A}, \mathbf{N}, \mathbf{b}, \mathbf{d}, \tau, k_{\max}$

// Initialization

$\beta_1 \mathbf{M}\mathbf{u}_1 = \mathbf{b}, \quad \alpha_1 \mathbf{N}\mathbf{v}_1 = \mathbf{A}^{\mathsf{T}}\mathbf{u}_1$ *// Initialize bidiagonalization*

$\delta_1 = 1, \quad \hat{\alpha}_1 = (\alpha_1^2 + 1)^{\frac{1}{2}}, \quad c_1 = \alpha_1/\hat{\alpha}_1, \quad s_1 = 1/\hat{\alpha}_1, \quad \zeta_1 = s_1\beta_1$

$\mathbf{d}_1 = s_1\mathbf{u}_1, \quad \mathbf{x}_1 = \zeta_1\mathbf{d}_1$

$k = 1, \quad \Delta = \zeta_1^2, \quad \text{converged} = \textbf{false}$

while not converged **and** $k < k_{\max}$ **do**

  // Continue bidiagonalization

  $\beta_{k+1}\mathbf{M}\mathbf{u}_{k+1} = \mathbf{A}\mathbf{v}_k - \alpha_k\mathbf{M}\mathbf{u}_k, \quad \alpha_{k+1}\mathbf{N}\mathbf{v}_{k+1} = \mathbf{A}^{\mathsf{T}}\mathbf{u}_{k+1} - \beta_{k+1}\mathbf{N}\mathbf{v}_k$

  $\hat{\beta}_{k+1} = c_k\beta_{k+1}, \quad \gamma_{k+1} = s_k\beta_{k+1}$ *// Continue reflection of type I*

  $\delta_{k+1} = (\gamma_{k+1}^2 + 1)^{\frac{1}{2}}, \ \bar{c}_k = -1/\delta_{k+1}, \ \bar{s}_k = \gamma_{k+1}/\delta_{k+1}$ *// Reflection of type II*

  // Compute next Givens reflection of type I

  $\hat{\alpha}_{k+1} = (\alpha_{k+1}^2 + \delta_{k+1}^2)^{\frac{1}{2}}, \quad c_{k+1} = \alpha_{k+1}/\hat{\alpha}_{k+1}, \quad s_{k+1} = \delta_{k+1}/\hat{\alpha}_{k+1}$

  $\zeta_{k+1} = -\hat{\beta}_{k+1}\zeta_k/\hat{\alpha}_{k+1}, \quad \Delta = \Delta + \zeta_{k+1}^2,$ *// Update*

  $\mathbf{d}_{k+1} = (\mathbf{u}_{k+1} - \hat{\beta}_{k+1}\mathbf{d}_k)/\hat{\alpha}_{k+1}$

  $\mathbf{x}_{k+1} = \mathbf{x}_k + \zeta_{k+1}\mathbf{d}_{k+1}$

  if $k \geq d - 1$ **then**

    converged $= \left(\sum_{j=k-d+2}^{k+1} \zeta_j^2 < \tau^2\Delta\right)$ *// Test convergence in* $\mathbf{x}$

  end if

  $k \leftarrow k + 1$

end while

$\mathbf{x} = \mathbf{x}_{k+1}$

$\mathbf{y} = \mathbf{N}^{-1}\mathbf{A}^{\mathsf{T}}\mathbf{x}$

return $(\mathbf{x}, \mathbf{y})$

and

$$\mathbf{A}^{\mathsf{T}}\mathbf{M}^{-1}(\mathbf{b} - \mathbf{A}\mathbf{y}) = (\mathbf{N}\mathbf{V}_k\mathbf{E}_k^{\mathsf{T}} + \alpha_{k+1}\mathbf{N}\mathbf{v}_{k+1}\mathbf{e}_{k+1}^{\mathsf{T}})(\beta_1\mathbf{e}_1 - \mathbf{E}_k\bar{\mathbf{y}}_k).$$

The identities

$$\mathbf{e}_{k+1}^{\mathsf{T}}\mathbf{e}_1 = 0, \quad \mathbf{e}_{k+1}^{\mathsf{T}}\mathbf{E}_k = \beta_{k+1}\mathbf{e}_k^{\mathsf{T}}, \quad \text{and} \quad \mathbf{E}_k^{\mathsf{T}}\mathbf{e}_1 = \alpha_1\mathbf{e}_1$$

simplify the above to

$$\mathbf{A}^{\mathsf{T}}\mathbf{M}^{-1}(\mathbf{b} - \mathbf{A}\mathbf{y}) = \mathbf{N}\mathbf{V}_{k+1}\left(\alpha_1\beta_1\mathbf{e}_1 - \begin{bmatrix} \mathbf{E}_k^{\mathsf{T}}\mathbf{E}_k \\ \alpha_{k+1}\beta_{k+1}\mathbf{e}_k^{\mathsf{T}} \end{bmatrix}\bar{\mathbf{y}}_k\right).$$

Similarly,

$$\mathbf{N}\mathbf{y} = \mathbf{N}\mathbf{V}_k\bar{\mathbf{y}}_k = \mathbf{N}\mathbf{V}_{k+1}\begin{bmatrix} \bar{\mathbf{y}}_k \\ 0 \end{bmatrix}.$$

By orthogonality of $\mathbf{N}^{\frac{1}{2}}\mathbf{V}_{k+1}$,

$$\begin{aligned}
&\|\mathbf{N}^{-\frac{1}{2}}(\mathbf{A}^\mathsf{T}\mathbf{M}^{-1}\mathbf{b}-(\mathbf{A}^\mathsf{T}\mathbf{M}^{-1}\mathbf{A}+\mathbf{N})\mathbf{y}))\|_2 \\
&\qquad = \left\|\mathbf{N}^{\frac{1}{2}}\mathbf{V}_{k+1}\left(\alpha_1\beta_1\mathbf{e}_1-\begin{bmatrix}\mathbf{E}_k^\mathsf{T}\mathbf{E}_k+\mathbf{I}_k\\ \alpha_{k+1}\beta_{k+1}\mathbf{e}_k^\mathsf{T}\end{bmatrix}\bar{\mathbf{y}}\right)\right\|_2 \\
&\qquad = \left\|\alpha_1\beta_1\mathbf{e}_1-\begin{bmatrix}\mathbf{E}_k^\mathsf{T}\mathbf{E}_k+\mathbf{I}_k\\ \alpha_{k+1}\beta_{k+1}\mathbf{e}_k^\mathsf{T}\end{bmatrix}\bar{\mathbf{y}}\right\|_2 .
\end{aligned} \tag{5.37}$$

Minimizing the latter residual is precisely the subproblem solved by the classic regularized LSMR with parameter $\lambda=1$. Once again, changing the Golub–Kahan procedure in LSMR is all that is required to solve (1.5) with $\mathbf{f}=\mathbf{b}$ and $\mathbf{g}=0$.

A by-product of the above is the underlying Lanczos process

$$\begin{aligned}
(\mathbf{N}+\mathbf{A}^\mathsf{T}\mathbf{M}^{-1}\mathbf{A})\mathbf{V}_{k+1} &= \mathbf{N}\big(\mathbf{V}_k(\mathbf{E}_k^\mathsf{T}\mathbf{E}_k+\mathbf{I}_k)+\alpha_{k+1}\beta_{k+1}\mathbf{v}_{k+1}\mathbf{e}_k^\mathsf{T}\big) \\
&= \mathbf{N}\mathbf{V}_{k+1}\begin{bmatrix}\mathbf{E}_k^\mathsf{T}\mathbf{E}_k+\mathbf{I}_k\\ \alpha_{k+1}\beta_{k+1}\mathbf{e}_k^\mathsf{T}\end{bmatrix},
\end{aligned} \tag{5.38}$$

which we already discovered in the proof of Theorem 5.1 and which is equivalent to what would be generated by applying the symmetric Lanczos process to $(\mathbf{N}+\mathbf{A}^\mathsf{T}\mathbf{M}^{-1}\mathbf{A})$ with initial vector $\mathbf{A}^\mathsf{T}\mathbf{M}^{-1}\mathbf{b}$ in the metric defined by $\mathbf{N}$. We detail the Lanczos process that characterizes MINRES on this set of normal equations in Section 6.4.

We have established the following result, which follows directly from the very definition of LSMR and the Lanczos process (5.38).

Theorem 5.5 (Relation between LSMR and MINRES). *The generalized* LSMR *iterates on* (2.4) *are the same as those generated by* MINRES *on the positive-definite system* (2.6) *in the metric defined by* $\mathbf{N}$.

5.6 ▪ Generalized LSMR: Implementation Details

Most of the details in this section come directly from Fong and Saunders (2011) but will turn out to be useful in designing a stopping condition in the appropriate norm.

The core of G-LSMR revolves around two QR factorizations. The first is the factorization of Chapter 5, i.e.,

$$\mathbf{Q}_k^\mathsf{T}\begin{bmatrix}\mathbf{E}_k\\ \mathbf{I}_k\end{bmatrix}=\begin{bmatrix}\mathbf{R}_k\\ 0\end{bmatrix}, \qquad \mathbf{R}_k=\begin{bmatrix}\rho_1 & \theta_2 & & \\ & \rho_2 & \ddots & \\ & & \ddots & \theta_k \\ & & & \rho_k\end{bmatrix}.$$

As before, $\mathbf{E}_k^\mathsf{T}\mathbf{E}_k+\mathbf{I}_k=\mathbf{R}_k^\mathsf{T}\mathbf{R}_k$. The least-squares problem (5.37) then reads

$$\underset{\bar{\mathbf{y}}}{\text{minimize}}\ \tfrac{1}{2}\left\|\alpha_1\beta_1\mathbf{e}_1-\begin{bmatrix}\mathbf{R}_k^\mathsf{T}\mathbf{R}_k\\ \alpha_{k+1}\beta_{k+1}\mathbf{e}_k^\mathsf{T}\end{bmatrix}\bar{\mathbf{y}}\right\|_2 . \tag{5.39}$$

Define the approximation $\mathbf{y} \approx \mathbf{y}_k := \mathbf{V}_k \bar{\mathbf{y}}_k$, where $\bar{\mathbf{y}}_k$ solves the above least-squares problem and let $\mathbf{t}_k := \mathbf{R}_k \bar{\mathbf{y}}_k$. Define also $\mathbf{q}_k := \mathbf{R}_k^{-\mathsf{T}}(\alpha_{k+1}\beta_{k+1}\mathbf{e}_k) = \gamma_k \mathbf{e}_k$, where

$$\gamma_k := \alpha_{k+1}\beta_{k+1}/\rho_k. \tag{5.40}$$

We now perform a second QR factorization

$$\bar{\mathbf{Q}}_k \begin{bmatrix} \mathbf{R}_k^{\mathsf{T}} & \alpha_1\beta_1\mathbf{e}_1 \\ \gamma_k \mathbf{e}_k^{\mathsf{T}} & 0 \end{bmatrix} = \begin{bmatrix} \bar{\mathbf{R}}_k & \mathbf{z}_k \\ 0 & \zeta_{k+1} \end{bmatrix}, \qquad \bar{\mathbf{R}}_k = \begin{bmatrix} \bar{\rho}_1 & \bar{\theta}_2 & & \\ & \bar{\rho}_2 & \ddots & \\ & & \ddots & \bar{\theta}_k \\ & & & \bar{\rho}_k \end{bmatrix}.$$

For this second factorization to be well defined recursively, it is necessary to show that $\gamma_k = \theta_{k+1}$ so that the matrix $\begin{bmatrix} \mathbf{R}_k & \gamma_k \mathbf{e}_k \end{bmatrix}$ is $\mathbf{R}_k$ with one extra column taken from $\mathbf{R}_{k+1}$. But upon examination of the sets of two reflections defining the first factorization in Chapter 5, we see that $\theta_{j+1} = \bar{s}_k \alpha_{j+1} = \alpha_{j+1}\beta_{j+1}/\rho_j = \gamma_j$. The least-squares problem in $\bar{\mathbf{y}}$ may now be restated as

$$\underset{\mathbf{t}}{\text{minimize}} \ \tfrac{1}{2} \left\| \begin{bmatrix} \mathbf{z}_k \\ \zeta_{k+1} \end{bmatrix} - \begin{bmatrix} \bar{\mathbf{R}}_k \\ 0 \end{bmatrix} \mathbf{t} \right\|_2.$$

The solution is obtained by solving $\bar{\mathbf{R}}_k \mathbf{t}_k := \mathbf{z}_k$ and the value of the residual is $|\zeta_{k+1}|$. Since $\bar{\mathbf{R}}_k$ is upper triangular, it is likely that all components of $\mathbf{t}_k$ change from one iteration to the next, and it is more efficient to seek an update of $\mathbf{y}_k$ directly. The first step to achieve this is to define the matrix $\mathbf{D}_k$ as in (5.6) and to additionally define $\bar{\mathbf{D}}_k$ from the solution of

$$\bar{\mathbf{R}}_k^{\mathsf{T}} \bar{\mathbf{D}}_k^{\mathsf{T}} = \mathbf{D}_k^{\mathsf{T}}. \tag{5.41}$$

This defines the columns $\bar{\mathbf{d}}_i$ of $\bar{\mathbf{D}}_k$ recursively as the solution of a lower triangular system:

$$\bar{\mathbf{d}}_0 := 0 \quad \text{and} \quad \bar{\mathbf{d}}_{i+1} := \frac{1}{\bar{\rho}_{i+1}}(\mathbf{d}_{i+1} - \bar{\theta}_{i+1}\bar{\mathbf{d}}_i), \quad i \geq 0. \tag{5.42}$$

With these definitions, we may update $\mathbf{y}_k$ as

$$\mathbf{y}_k = \mathbf{V}_k \bar{\mathbf{y}}_k = \mathbf{V}_k \mathbf{R}_k^{-1} \mathbf{t}_k = \mathbf{D}_k \mathbf{t}_k = \bar{\mathbf{D}}_k \bar{\mathbf{R}}_k \mathbf{t}_k = \bar{\mathbf{D}}_k \mathbf{z}_k = \mathbf{y}_{k-1} + \zeta_k \bar{\mathbf{d}}_k. \tag{5.43}$$

The matrix $\bar{\mathbf{D}}_k$ yields a partial factorization of the operator that determines the energy norm in which errors should be measured in G-LSMR. The rationale behind this energy norm is the following. Consider a hypothetical symmetric and positive-definite system $\mathbf{C}\mathbf{x} = \mathbf{b}$. It is the defining property of MINRES that $\|\mathbf{r}_k\|_2$ decreases monotonically, where $\mathbf{r}_k := \mathbf{b} - \mathbf{C}\mathbf{x}_k = \mathbf{C}(\mathbf{x}_* - \mathbf{x}_k)$ and $\mathbf{x}_*$ is the unique solution of the system. Thus, $\|\mathbf{r}_k\|_2 = \|\mathbf{x}_* - \mathbf{x}_k\|_{\mathbf{C}^2}$ is the quantity that decreases. Suppose now, as in (5.36), that the residual must be measured in the $\mathbf{N}^{-1}$-norm. Then

$$\|\mathbf{r}_k\|_{\mathbf{N}^{-1}} = \|\mathbf{N}^{-\frac{1}{2}}\mathbf{r}_k\|_2 = \|\mathbf{N}^{-\frac{1}{2}}\mathbf{C}(\mathbf{x}_* - \mathbf{x}_k)\| = \|\mathbf{x}_* - \mathbf{x}_k\|_{\mathbf{C}\mathbf{N}^{-1}\mathbf{C}}$$

is the appropriate energy norm in which the error should be measured in MINRES. The next result summarizes the main properties in our SQD context.

Theorem 5.6 (Solution and Error Norm in LSMR). *Let* $\overline{\mathbf{D}}_k$ *be defined as in* (5.41). *Then, for* $k = 1, \ldots, m$, *we have*

$$\overline{\mathbf{D}}_k^{\mathsf{T}} \mathbf{G} \overline{\mathbf{D}}_k = \mathbf{I}_k, \quad \textit{where } \mathbf{G} := (\mathbf{A}^{\mathsf{T}}\mathbf{M}^{-1}\mathbf{A}+\mathbf{N})\mathbf{N}^{-1}(\mathbf{A}^{\mathsf{T}}\mathbf{M}^{-1}\mathbf{A}+\mathbf{N}). \tag{5.44}$$

In particular,

$$\mathbf{y}_k = \sum_{j=1}^{k} \zeta_j \overline{\mathbf{d}}_j \tag{5.45}$$

and

$$\|\mathbf{y}_k\|_{\mathbf{G}}^2 = \sum_{j=1}^{k} \zeta_j^2, \tag{5.46}$$

and we have the error estimate

$$\|\mathbf{y}-\mathbf{y}_k\|_{\mathbf{G}}^2 = \sum_{j=k+1}^{m} \zeta_j^2, \tag{5.47}$$

where $\mathbf{y}$ *is the solution of* (5.36).

Proof. We establish (5.44). The expansion (5.45) follows by repeated application of (5.43). The proof of (5.46) and (5.47) is a direct consequence of (5.44) and (5.45), as in the proof of Theorem 5.4.

We deduce from (5.41) and (5.6) that $\overline{\mathbf{D}}_k^{\mathsf{T}}\mathbf{G}\overline{\mathbf{D}}_k = \overline{\mathbf{R}}_k^{-\mathsf{T}}\mathbf{R}_k^{-\mathsf{T}}\mathbf{V}_k^{\mathsf{T}}\mathbf{G}\mathbf{V}_k\mathbf{R}_k^{-1}\overline{\mathbf{R}}_k^{-1}$. We now expand this expression from the inside and out. Using (4.3), we have

$$\begin{aligned}(\mathbf{A}^{\mathsf{T}}\mathbf{M}^{-1}\mathbf{A}+\mathbf{N})\mathbf{V}_k &= \mathbf{A}^{\mathsf{T}}\mathbf{U}_{k+1}\mathbf{E}_k + \mathbf{N}\mathbf{V}_k \\ &= \mathbf{N}((\mathbf{V}_k\mathbf{E}_k^{\mathsf{T}} + \alpha_{k+1}\mathbf{v}_{k+1}\mathbf{e}_{k+1}^{\mathsf{T}})\mathbf{E}_k + \mathbf{V}_k) \\ &= \mathbf{N}(\mathbf{V}_k(\mathbf{E}_k^{\mathsf{T}}\mathbf{E}_k + \mathbf{I}_k) + \gamma_k\rho_k\mathbf{v}_{k+1}\mathbf{e}_k^{\mathsf{T}}),\end{aligned}$$

where we used the identity $\mathbf{e}_{k+1}^{\mathsf{T}}\mathbf{E}_k = \beta_{k+1}\mathbf{e}_k^{\mathsf{T}}$ and the definition (5.40) of γ_k. Using the above identity twice, we obtain, after some basic manipulations,

$$\mathbf{V}_k^{\mathsf{T}}\mathbf{G}\mathbf{V}_k = (\mathbf{E}_k^{\mathsf{T}}\mathbf{E}_k + \mathbf{I}_k)^2 + \gamma_k^2\rho_k^2\mathbf{e}_k\mathbf{e}_k^{\mathsf{T}}.$$

In the product above, cross terms vanish because they contain the expression $\mathbf{V}_k^{\mathsf{T}}\mathbf{N}\mathbf{v}_{k+1}$, which is zero by orthogonality.

As we already noticed at the beginning of this section, $\mathbf{E}_k^{\mathsf{T}}\mathbf{E}_k + \mathbf{I}_k = \mathbf{R}_k^{\mathsf{T}}\mathbf{R}_k$, and therefore

$$(\mathbf{E}_k^{\mathsf{T}}\mathbf{E}_k + \mathbf{I}_k)^2 = \mathbf{R}_k^{\mathsf{T}}\mathbf{R}_k\mathbf{R}_k^{\mathsf{T}}\mathbf{R}_k.$$

Consequently,

$$\mathbf{R}_k^{-\mathsf{T}}\left((\mathbf{E}_k^{\mathsf{T}}\mathbf{E}_k + \mathbf{I}_k)^2 + \gamma_k^2\rho_k^2\mathbf{e}_k\mathbf{e}_k^{\mathsf{T}}\right)\mathbf{R}_k^{-1} = \mathbf{R}_k\mathbf{R}_k^{\mathsf{T}} + \gamma_k^2\mathbf{e}_k\mathbf{e}_k^{\mathsf{T}},$$

where we used the identity $\mathbf{R}_k^{-\mathsf{T}}\mathbf{e}_k = \rho_k^{-1}\mathbf{e}_k$.

Note now that by definition of $\overline{\mathbf{R}}_k$,

$$\overline{\mathbf{R}}_k^{\mathsf{T}}\overline{\mathbf{R}}_k = \begin{bmatrix}\overline{\mathbf{R}}_k^{\mathsf{T}} & 0^{\mathsf{T}}\end{bmatrix}\begin{bmatrix}\overline{\mathbf{R}}_k \\ 0\end{bmatrix} = \begin{bmatrix}\mathbf{R}_k & \gamma_k\mathbf{e}_k\end{bmatrix}\overline{\mathbf{Q}}_k\overline{\mathbf{Q}}_k^{\mathsf{T}}\begin{bmatrix}\mathbf{R}_k^{\mathsf{T}} \\ \gamma_k\mathbf{e}_k^{\mathsf{T}}\end{bmatrix} = \mathbf{R}_k\mathbf{R}_k^{\mathsf{T}} + \gamma_k^2\mathbf{e}_k\mathbf{e}_k^{\mathsf{T}}.$$

This last identity finally yields

$$\bar{\mathbf{D}}_k^\mathsf{T} \mathbf{G} \bar{\mathbf{D}}_k = \bar{\mathbf{R}}_k^{-\mathsf{T}} (\mathbf{R}_k \mathbf{R}_k^\mathsf{T} + \gamma_k^2 \mathbf{e}_k \mathbf{e}_k^\mathsf{T}) \bar{\mathbf{R}}_k^{-1} = \bar{\mathbf{R}}_k^{-\mathsf{T}} (\bar{\mathbf{R}}_k^\mathsf{T} \bar{\mathbf{R}}_k) \bar{\mathbf{R}}_k^{-1} = \mathbf{I}_k,$$

and this completes the proof. □

The last reflection computes $\bar{\mathbf{R}}_k$ by eliminating the subdiagonals of $\mathbf{R}_k^\mathsf{T}$, i.e., θ_{j+1} for $j = 1, \dots, k-1$, as follows:

$$\begin{matrix} & k & k+1 \\ k \\ k+1 \end{matrix} \begin{bmatrix} \hat{c}_k & \hat{s}_k \\ \hat{s}_k & -\hat{c}_k \end{bmatrix} \begin{bmatrix} \tilde{\rho}_k & \\ \theta_{k+1} & \rho_{k+1} \end{bmatrix} = \begin{bmatrix} \bar{\rho}_k & \bar{\theta}_{k+1} \\ & \tilde{\rho}_{k+1} \end{bmatrix},$$

with the initialization $\tilde{\rho}_1 := \rho_1$. In other words, $\bar{\rho}_k := \sqrt{\tilde{\rho}_k^2 + \theta_{k+1}^2}$, $\hat{c}_k := \tilde{\rho}_k / \bar{\rho}_k$, $\hat{s}_k := \theta_{k+1} / \bar{\rho}_k$, $\bar{\theta}_{k+1} := \hat{s}_k \rho_{k+1}$, and $\tilde{\rho}_{k+1} := -\hat{c}_k \rho_{k+1}$.

During the kth iteration, ρ_{k+1} is not yet available. Therefore, the computation of $\bar{\theta}_{k+1}$ and $\tilde{\rho}_{k+1}$ can take place only during the next iteration. The kth iteration computes $\bar{\theta}_k$ and $\tilde{\rho}_k$ as soon as ρ_k becomes available using the previous values $\hat{c}_{k-1}$ and $\hat{s}_{k-1}$. To update $\mathbf{y}_k$, we first compute $\bar{\mathbf{d}}_k$ according to (5.42) and use (5.43). Computing $\mathbf{d}_{k+1}$ appears to require ρ_{k+1}. However, as Fong and Saunders (2011) point out, it is possible to bypass this requirement by defining

$$\mathbf{h}_k := \rho_k \mathbf{d}_k \quad \text{and} \quad \bar{\mathbf{h}}_k := \rho_k \bar{\rho}_k \bar{\mathbf{d}}_k$$

and updating

$$\bar{\mathbf{h}}_k = \mathbf{h}_k + \frac{\rho_k \bar{\theta}_k}{\rho_{k-1} \bar{\rho}_{k-1}} \bar{\mathbf{h}}_{k-1}, \quad \mathbf{y}_k = \mathbf{y}_{k-1} + \frac{\zeta_k}{\rho_{k-1} \bar{\rho}_{k-1}} \bar{\mathbf{h}}_k, \quad \mathbf{h}_{k+1} = \mathbf{v}_{k+1} - \frac{\theta_{k+1}}{\rho_k} \mathbf{h}_k.$$

The least-squares residual is initially given by

$$\begin{bmatrix} \hat{c}_1 & \hat{s}_1 \\ \hat{s}_1 & -\hat{c}_1 \end{bmatrix} \begin{bmatrix} \beta_1 \\ 0 \end{bmatrix} = \begin{bmatrix} \hat{c}_1 \beta_1 \\ \hat{s}_1 \beta_1 \end{bmatrix},$$

so that $\zeta_1 := \hat{c}_1 \beta_1$. Define $\hat{\zeta}_2 := \hat{s}_1 \beta_1$ to obtain the recursion

$$\begin{bmatrix} \hat{c}_k & \hat{s}_k \\ \hat{s}_k & -\hat{c}_k \end{bmatrix} \begin{bmatrix} \hat{\zeta}_k \\ 0 \end{bmatrix} = \begin{bmatrix} \hat{c}_k \hat{\zeta}_k \\ \hat{s}_k \hat{\zeta}_k \end{bmatrix},$$

i.e., $\zeta_k := \hat{c}_k \hat{\zeta}_k$ and $\hat{\zeta}_{k+1} := \hat{s}_k \hat{\zeta}_k$. The recursion begins with $\hat{\zeta}_1 := \zeta_1$. Note that there is a lag in the residual value—the component $\hat{\zeta}_{k+1}$ is obtained when computing $\mathbf{x}_k$, but it is $|\zeta_k|$ that gives the residual value, and it corresponds to $\mathbf{x}_{k-1}$, not to $\mathbf{x}_k$. The main computational steps of G-LSMR are summarized as Algorithm 5.3.

5.7 ▪ Generalized CRAIG-MR

In this section, we present the main features of a method that is similar to LSMR but that is equivalent to applying MINRES to the Schur-complement equations. By analogy with LSMR, we dub this method CRAIG-MR. Its application to (5.16) is dubbed

Algorithm 5.3 Generalized LSMR

Require: $\mathbf{M}, \mathbf{A}, \mathbf{N}, \mathbf{b}, d, \tau, k_{max}$

$\beta_1 \mathbf{M}\mathbf{u}_1 = \mathbf{b}, \quad \alpha_1 \mathbf{N}\mathbf{v}_1 = \mathbf{A}^\mathsf{T}\mathbf{u}_1$ // *Initialize bidiagonalization*

$\hat{\alpha}_1 = \alpha_1, \quad \bar{\rho}_1 = \alpha_1, \quad \hat{\zeta}_1 = \alpha_1\beta_1, \quad \hat{c}_0 = -1, \quad \hat{s}_0 = 0, \quad \rho_0 = 0, \quad \bar{\rho}_0 = 0$

$\mathbf{h}_1 = \mathbf{v}_1, \quad \bar{\mathbf{h}}_0 = 0, \quad \mathbf{y}_0 = 0$

$k = 1, \quad \Delta = 0,$ converged = **false**

while not converged **and** $k < k_{max}$ **do**

  // *Continue bidiagonalization*

  $\beta_{k+1}\mathbf{M}\mathbf{u}_{k+1} = \mathbf{A}\mathbf{v}_k - \alpha_k \mathbf{M}\mathbf{u}_k, \quad \alpha_{k+1}\mathbf{N}\mathbf{v}_{k+1} = \mathbf{A}^\mathsf{T}\mathbf{u}_{k+1} - \beta_{k+1}\mathbf{N}\mathbf{v}_k$

  $\hat{\rho}_k = (1+\hat{\alpha}_k^2)^{\frac{1}{2}}, \quad c_k = \hat{\alpha}_k/\hat{\rho}_k, \quad s_k = 1/\hat{\rho}_k$ // *Reflection of type II*

  $\rho_k = (\hat{\rho}_k^2 + \beta_{k+1}^2)^{\frac{1}{2}}, \quad \bar{c}_k = \hat{\rho}_k/\rho_k, \quad \bar{s}_k = \beta_{k+1}/\rho_k$ // *Reflection of type I*

  $\theta_{k+1} = \bar{s}_k \alpha_{k+1}, \quad \hat{\alpha}_{k+1} = -\bar{c}_k \alpha_{k+1},$

  $\bar{\theta}_k = \hat{s}_{k-1}\rho_k, \quad \tilde{\rho}_k = -\hat{c}_{k-1}\rho_k$ // *Reflection of type III*

  $\bar{\rho}_k = (\tilde{\rho}_k^2 + \theta_{k+1}^2)^{\frac{1}{2}}, \quad \hat{c}_k = \tilde{\rho}_k/\bar{\rho}_k, \quad \hat{s}_k = \theta_{k+1}/\bar{\rho}_k,$

  $\zeta_k = \hat{c}_k\hat{\zeta}_k, \quad \hat{\zeta}_{k+1} = \hat{s}_k\hat{\zeta}_k, \quad \Delta = \Delta + \zeta_k^2$ // *Residual and error update*

  $\bar{\mathbf{h}}_k = \mathbf{h}_k + \rho_k\bar{\theta}_k/(\rho_{k-1}\bar{\rho}_{k-1})\bar{\mathbf{h}}_{k-1}$ // *Update*

  $\mathbf{y}_k = \mathbf{y}_{k-1} + \zeta_k/(\rho_{k-1}\bar{\rho}_{k-1})\bar{\mathbf{h}}_k$

  $\mathbf{h}_{k+1} = \mathbf{v}_{k+1} - \theta_{k+1}/\rho_k \mathbf{h}_k$

  if $k \geq d$ **then**

    converged $= \left(\sum_{j=k-d+1}^{k} \zeta_j^2 < \tau^2\Delta\right)$ // *Test convergence*

  end if

  $k \leftarrow k+1$

end while

$\mathbf{y} = \mathbf{y}_k$

$\mathbf{x} = \mathbf{M}^{-1}(\mathbf{b} - \mathbf{A}\mathbf{y})$

return $(\mathbf{x}, \mathbf{y})$

the *generalized* CRAIG-MR, or G-CRAIG-MR. The reason for introducing this method becomes clear in Section 6.4, where we show that MINRES applied directly to (1.5) with right-hand side $(\mathbf{b}, 0)$ alternates between G-LSMR steps and G-CRAIG-MR steps.

It follows from (5.19) that the approximation $\mathbf{x}_k = \mathbf{U}_k\bar{\mathbf{x}}_k$ reveals the associated Lanczos process

$$\begin{aligned}(\mathbf{M}+\mathbf{A}\mathbf{N}^{-1}\mathbf{A}^\mathsf{T})\mathbf{U}_k &= \mathbf{M}\mathbf{U}_k(\mathbf{B}_k\mathbf{B}_k^\mathsf{T} + \mathbf{I}_k) + \alpha_k\beta_{k+1}\mathbf{M}\mathbf{u}_{k+1}\mathbf{e}_k^\mathsf{T} \\ &= \mathbf{M}\mathbf{U}_{k+1}\begin{bmatrix}\mathbf{B}_k\mathbf{B}_k^\mathsf{T} + \mathbf{I}_k \\ \alpha_k\beta_{k+1}\mathbf{e}_k^\mathsf{T}\end{bmatrix},\end{aligned} \tag{5.48}$$

where we note that $(\mathbf{B}_k\mathbf{B}_k^\mathsf{T} + \mathbf{I}_k)$ is tridiagonal and, by definition, $\mathbf{b} = \beta_1\mathbf{M}\mathbf{u}_1$.

Note that for any $\mathbf{x}$, the residual of (5.16) lies in $\mathbb{M}^\star$. The appropriate norm to measure this residual is thus the $\mathbf{M}^{-1}$-norm. By definition, G-CRAIG-MR computes $\mathbf{x}_k$ as the solution of

$$\underset{\mathbf{x}\in\mathrm{Range}(\mathbf{V}_k)}{\text{minimize}} \ \tfrac{1}{2}\|\mathbf{M}^{-\frac{1}{2}}(\mathbf{b} - (\mathbf{M}+\mathbf{A}\mathbf{N}^{-1}\mathbf{A}^\mathsf{T})\mathbf{x})\|_2.$$

Since the system is always consistent in $\mathbb{R}^n$, the final solution $\mathbf{x}$ solves

$$\mathbf{M}^{-\frac{1}{2}}(\mathbf{M}+\mathbf{A}\mathbf{N}^{-1}\mathbf{A}^\mathsf{T})\mathbf{x}=\mathbf{M}^{-\frac{1}{2}}\mathbf{b}. \tag{5.49}$$

We have established the following result, which follows directly from the very definition of G-CRAIG-MR and the Lanczos process (5.48).

Theorem 5.7 (Relation between CRAIG-MR and MINRES). *The generalized* CRAIG-MR *iterates on* (2.4) *are the same as those generated by the* MINRES *method on the positive-definite system* (2.7) *in the metric defined by* **M**.

5.8 ▪ Generalized CRAIG-MR: Implementation Details

If we use the approximation $\mathbf{x} \approx \mathbf{x}_k = \mathbf{U}_k\bar{\mathbf{x}}_k$ and (5.48), the kth subproblem may be written as

$$\underset{\bar{\mathbf{x}}_k}{\text{minimize}}\ \tfrac{1}{2}\left\|\beta_1\mathbf{e}_1-\begin{bmatrix}\mathbf{B}_k\mathbf{B}_k^\mathsf{T}+\mathbf{I}_k\\ \alpha_k\beta_{k+1}\mathbf{e}_k^\mathsf{T}\end{bmatrix}\bar{\mathbf{x}}_k\right\|_2, \tag{5.50}$$

which is again a regularized subproblem with regularization parameter $\lambda=1$.

An implementation of G-CRAIG-MR starts, as in G-CRAIG, by computing the LQ factorization (5.14), where $\hat{\mathbf{B}}_k$ is $k\times k$ lower bidiagonal. As before, $\mathbf{B}_k\mathbf{B}_k^\mathsf{T}+\mathbf{I}_k=\hat{\mathbf{B}}_k\hat{\mathbf{B}}_k^\mathsf{T}$. We define

$$\mathbf{t}_k := \hat{\mathbf{B}}_k^\mathsf{T}\bar{\mathbf{x}}_k \quad\text{and}\quad \mathbf{q}_k := \hat{\mathbf{B}}_k^{-1}(\alpha_k\beta_{k+1}\mathbf{e}_k)=\frac{\alpha_k\beta_{k+1}}{\hat{\alpha}_k}\mathbf{e}_k := \gamma_k\mathbf{e}_k.$$

The least-squares residual of the kth subproblem may now be rewritten as

$$\beta_1\mathbf{e}_1-\begin{bmatrix}\mathbf{B}_k\mathbf{B}_k^\mathsf{T}+\mathbf{I}_k\\ \alpha_k\beta_{k+1}\mathbf{e}_k^\mathsf{T}\end{bmatrix}\bar{\mathbf{x}}_k=\beta_1\mathbf{e}_1-\begin{bmatrix}\hat{\mathbf{B}}_k\hat{\mathbf{B}}_k^\mathsf{T}\\ \mathbf{q}_k^\mathsf{T}\hat{\mathbf{B}}_k^\mathsf{T}\end{bmatrix}\bar{\mathbf{x}}_k=\beta_1\mathbf{e}_1-\begin{bmatrix}\hat{\mathbf{B}}_k\\ \gamma_k\mathbf{e}_k^\mathsf{T}\end{bmatrix}\mathbf{t}_k.$$

As in LSMR, we now perform the third QR factorization

$$\bar{\mathbf{Q}}_{k+1}\begin{bmatrix}\hat{\mathbf{B}}_k & \beta_1\mathbf{e}_1\\ \gamma_k\mathbf{e}_k^\mathsf{T} & 0\end{bmatrix}=\begin{bmatrix}\mathbf{R}_k & \mathbf{z}_k\\ 0 & \zeta_{k+1}\end{bmatrix}\quad\text{with}\quad \mathbf{R}_k=\begin{bmatrix}\rho_1 & \theta_2 & & \\ & \rho_2 & \ddots & \\ & & \ddots & \theta_k\\ & & & \rho_k\end{bmatrix},$$

where $\bar{\mathbf{Q}}_{k+1}$ is orthogonal. Again, for this second factorization to be well defined recursively, it is necessary to verify that $\gamma_k=\hat{\beta}_{k+1}$. Fortunately, we have already verified this fact in (5.24). This helps us write the kth subproblem as

$$\underset{\mathbf{t}_k}{\text{minimize}}\ \tfrac{1}{2}\left\|\begin{bmatrix}\mathbf{z}_k\\ \zeta_{k+1}\end{bmatrix}-\begin{bmatrix}\mathbf{R}_k\\ 0\end{bmatrix}\mathbf{t}_k\right\|_2.$$

Clearly, we obtain the solution by solving $\mathbf{R}_k\mathbf{t}_k := \mathbf{z}_k$, and the least-squares residual is simply $|\zeta_{k+1}|$. The remaining algorithmic details essentially follow those developed by Fong and Saunders (2011) for LSMR.

Seeking once again a direct update of $\mathbf{x}_k$, we define the columns $\mathbf{d}_1$ to $\mathbf{d}_k$ of $\mathbf{D}_k$ recursively via the definition

$$\hat{\mathbf{B}}_k \mathbf{D}_k^\mathsf{T} = \mathbf{U}_k^\mathsf{T},$$

which yields

$$\mathbf{d}_1 := \frac{1}{\hat{\alpha}_1}\mathbf{u}_1 \qquad \text{and} \qquad \mathbf{d}_{j+1} := \frac{1}{\hat{\alpha}_{j+1}}(\mathbf{u}_{j+1} - \hat{\beta}_{j+1}\mathbf{d}_j) \quad (j \geq 0).$$

Similarly, define the columns $\bar{\mathbf{d}}_1$ to $\bar{\mathbf{d}}_k$ of $\bar{\mathbf{D}}_k$ via

$$\mathbf{R}_k^\mathsf{T} \bar{\mathbf{D}}_k^\mathsf{T} = \mathbf{D}_k^\mathsf{T}, \tag{5.51}$$

which yields

$$\bar{\mathbf{d}}_1 := \frac{1}{\rho_1}\mathbf{d}_1 \qquad \text{and} \qquad \bar{\mathbf{d}}_{j+1} := \frac{1}{\rho_{j+1}}(\mathbf{d}_{j+1} - \theta_{j+1}\bar{\mathbf{d}}_j) \quad (j \geq 0).$$

With these definitions, we may use the update

$$\mathbf{x}_k = \mathbf{U}_k \bar{\mathbf{x}}_k = \mathbf{D}_k \hat{\mathbf{B}}_k^\mathsf{T} \bar{\mathbf{x}}_k = \mathbf{D}_k \mathbf{t}_k = \bar{\mathbf{D}}_k \mathbf{R}_k \mathbf{t}_k = \bar{\mathbf{D}}_k \mathbf{z}_k = \mathbf{x}_{k-1} + \zeta_k \bar{\mathbf{d}}_k.$$

Theorem 5.8 (Solution and Error Norm in CRAIG-MR). *Let $\bar{\mathbf{D}}_k$ be defined as in* (5.51). *Then, for $k = 1, \ldots, n$, we have*

$$\bar{\mathbf{D}}_k^\mathsf{T} \mathbf{W} \bar{\mathbf{D}}_k = \mathbf{I}_k, \quad \textit{where } \mathbf{W} := (\mathbf{A}\mathbf{N}^{-1}\mathbf{A}^\mathsf{T} + \mathbf{M})\mathbf{M}^{-1}(\mathbf{A}\mathbf{N}^{-1}\mathbf{A}^\mathsf{T} + \mathbf{M}). \tag{5.52}$$

In particular,

$$\mathbf{x}_k = \sum_{j=1}^{k} \zeta_j \bar{\mathbf{d}}_j \tag{5.53}$$

and

$$\|\mathbf{x}_k\|_\mathbf{W}^2 = \sum_{j=1}^{k} \zeta_j^2, \tag{5.54}$$

and we have the error estimate

$$\|\mathbf{x} - \mathbf{x}_k\|_\mathbf{W}^2 = \sum_{j=k+1}^{n} \zeta_j^2, \tag{5.55}$$

where $\mathbf{x}$ *is the solution of* (5.49).

Proof. The proof is analogous to that of Theorem 5.6. □

The reflection necessary to compute $\mathbf{R}_k$, which we call a reflection of type III, eliminates the subdiagonals of $\hat{\mathbf{B}}_k$, i.e., $\hat{\beta}_{j+1}$, $j = 1, \ldots, k-1$, as follows:

$$\begin{array}{c} \\ k \\ k+1 \end{array} \overset{\begin{array}{cc} k & k+1 \end{array}}{\begin{bmatrix} \hat{c}_k & \hat{s}_k \\ \hat{s}_k & -\hat{c}_k \end{bmatrix}} \overset{\begin{array}{cc} k & k+1 \end{array}}{\begin{bmatrix} \tilde{\alpha}_k & \\ \hat{\beta}_{k+1} & \hat{\alpha}_{k+1} \end{bmatrix}} = \begin{bmatrix} \rho_k & \theta_{k+1} \\ & \tilde{\alpha}_{k+1} \end{bmatrix},$$

with the initialization $\widetilde{\alpha}_1 := \hat{\alpha}_1$. In other words, $\rho_k := \sqrt{\widetilde{\alpha}_k^2 + \hat{\beta}_{k+1}^2}$, $\hat{c}_k := \widetilde{\alpha}_k / \rho_k$, $\hat{s}_k := \hat{\beta}_{k+1}/\rho_k$, $\theta_{k+1} := \hat{s}_k \hat{\alpha}_{k+1}$, and $\widetilde{\alpha}_{k+1} = -\hat{c}_k \hat{\alpha}_{k+1}$.

The main details of G-CRAIG-MR are summarized as Algorithm 5.4. It is worth noting again that the least-squares residual, $|\zeta_k|$, lags one step behind and corresponds to the *previous* iterate $\mathbf{x}_{k-1}$.

Algorithm 5.4 Generalized CRAIG-MR

Require: $\mathbf{M}, \mathbf{A}, \mathbf{N}, \mathbf{b}, d, \tau, k_{\max}$

$\beta_1 \mathbf{M}\mathbf{u}_1 = \mathbf{b}, \quad \alpha_1 \mathbf{N}\mathbf{v}_1 = \mathbf{A}^T \mathbf{u}_1$ *// Initialize bidiagonalization*

$\delta_1 = 1, \quad \hat{\alpha}_1 = (\alpha_1^2 + 1)^{\frac{1}{2}}, \quad c_1 = \alpha_1/\hat{\alpha}_1, \quad s_1 = 1/\hat{\alpha}_1$

$\hat{\zeta}_1 = \beta_1, \quad \widetilde{\alpha}_1 = \hat{\alpha}_1, \quad \theta_1 = 0$

$\mathbf{d}_1 = 1/\hat{\alpha}_1 \mathbf{u}_1, \quad \bar{\mathbf{d}}_0 = 0, \quad \mathbf{x}_0 = 0$

$k = 1, \quad \Delta = 0, \quad$ converged $=$ **false**

while not converged **and** $k < k_{max}$ **do**

  // Continue bidiagonalization

  $\beta_{k+1} \mathbf{M}\mathbf{u}_{k+1} = \mathbf{A}\mathbf{v}_k - \alpha_k \mathbf{M}\mathbf{u}_k, \quad \alpha_{k+1} \mathbf{N}\mathbf{v}_{k+1} = \mathbf{A}^T \mathbf{u}_{k+1} - \beta_{k+1} \mathbf{N}\mathbf{v}_k$

  $\hat{\beta}_{k+1} = c_k \beta_{k+1}, \quad \gamma_{k+1} = s_k \beta_{k+1}$ *// Continue reflection of type I*

  $\delta_{k+1} = (\gamma_{k+1}^2 + 1)^{\frac{1}{2}}, \ \bar{c}_k = -1/\delta_{k+1}, \ \bar{s}_k = \gamma_{k+1}/\delta_{k+1}$ *// Reflection of type II*

  // Compute new Givens reflection of type I

  $\hat{\alpha}_{k+1} = (\alpha_{k+1}^2 + \delta_{k+1}^2)^{\frac{1}{2}}, \quad c_{k+1} = \alpha_{k+1}/\hat{\alpha}_{k+1}, \quad s_{k+1} = \delta_{k+1}/\hat{\alpha}_{k+1}$

  $\rho_k = (\widetilde{\alpha}_k^2 + \hat{\beta}_{k+1}^2)^{\frac{1}{2}}, \quad \hat{c}_k = \widetilde{\alpha}_k/\rho_k, \quad \hat{s}_k = \hat{\beta}_{k+1}/\rho_k$ *// Reflection of type III*

  $\theta_{k+1} = \hat{s}_k \hat{\alpha}_{k+1}, \quad \widetilde{\alpha}_{k+1} = -\hat{c}_k \hat{\alpha}_{k+1}$

  $\zeta_k = \hat{c}_k \hat{\zeta}_k, \quad \hat{\zeta}_{k+1} = \hat{s}_k \hat{\zeta}_k, \quad \Delta = \Delta + \zeta_k^2$ *// Update*

  $\mathbf{d}_{k+1} = 1/\hat{\alpha}_{k+1}(\mathbf{u}_{k+1} - \hat{\beta}_{k+1} \mathbf{d}_k)$

  $\bar{\mathbf{d}}_k = 1/\rho_k (\mathbf{d}_k - \theta_k \bar{\mathbf{d}}_{k-1})$

  $\mathbf{x}_k = \mathbf{x}_{k-1} + \zeta_k \bar{\mathbf{d}}_k$

  if $k \ge d$ **then**

    converged $= \left(\sum_{j=k-d+1}^{k} \zeta_j^2 < \tau^2 \Delta\right)$ *// Test convergence*

  end if

  $k \leftarrow k + 1$

end while

$\mathbf{x} = \mathbf{x}_k$

$\mathbf{y} = \mathbf{N}^{-1} \mathbf{A}^T \mathbf{x}$

return $(\mathbf{x}, \mathbf{y})$

5.9 ▪ Upper-Bound Error Estimates

In the previous sections, we proposed several lower bounds on the direct error that are linked to the techniques described by Golub and Meurant (2010). Even though those lower bounds estimate the error at step $k-d$, it is safe to use the most recent solution estimate, i.e., that computed at step k, owing to the monotonicity of the error sequence.

To motivate our approach, consider the generalized LSQR method and the relation (5.5), which determines the coefficients $\bar{\mathbf{y}}_k$ of the kth approximation $\mathbf{y}_k$ in terms of the initial values $\beta_1 = \|\mathbf{b}\|_{\mathbf{M}^{-1}}$ and $\alpha_1 = \|\mathbf{A}^T\mathbf{u}_1\|_{\mathbf{N}^{-1}} = \beta_1^{-1}\|\mathbf{A}^T\mathbf{M}^{-1}\mathbf{b}\|_{\mathbf{N}^{-1}}$. Let $\hat{\mathbf{T}}_k$ denote the symmetric and positive-definite tridiagonal matrix $\mathbf{R}_k^T\mathbf{R}_k$. Then $\bar{\mathbf{y}}_k = \alpha_1\beta_1\hat{\mathbf{T}}_k^{-1}\mathbf{e}_1$. Theorem 5.2 indicates that $\|\mathbf{y}_k\|_{\mathbf{A}^T\mathbf{M}^{-1}\mathbf{A}+\mathbf{N}} = \|\mathbf{z}_k\|_2$, where $\mathbf{z}_k = \mathbf{R}_k\bar{\mathbf{y}}_k$. Therefore,

$$\|\mathbf{y}_k\|_{\mathbf{A}^T\mathbf{M}^{-1}\mathbf{A}+\mathbf{N}} = \|\mathbf{z}_k\|_2 = \|\mathbf{R}_k\bar{\mathbf{y}}_k\|_2 = \|\bar{\mathbf{y}}_k\|_{\hat{\mathbf{T}}_k} = |\alpha_1\beta_1|\,\|\mathbf{e}_1\|_{\hat{\mathbf{T}}_k^{-1}}.$$

But $\|\mathbf{e}_1\|^2_{\hat{\mathbf{T}}_k^{-1}} = \mathbf{e}_1^T\hat{\mathbf{T}}_k^{-1}\mathbf{e}_1$, and so the above states that the squared energy norm of $\mathbf{y}_k$ is a factor of the leading element of $\hat{\mathbf{T}}_k^{-1}$. Using the same logic as in the proof of Theorem 5.2, if $\hat{\mathbf{T}} := \hat{\mathbf{T}}_n$ is the value of the tridiagonal when convergence has occurred, then we may measure the direct error as

$$\|\mathbf{y}_k - \mathbf{y}\|_{\mathbf{A}^T\mathbf{M}^{-1}\mathbf{A}+\mathbf{N}} = |\alpha_1\beta_1|\,\left|\|\mathbf{e}_1\|_{\mathbf{T}_k^{-1}} - \|\mathbf{e}_1\|_{\mathbf{T}^{-1}}\right|,$$

and this direct error is related to the quality of the approximation of the leading entry of $\hat{\mathbf{T}}^{-1}$ at step k. The same reasoning holds for the generalized LSMR method using (5.39). A similar result follows for the generalized CRAIG and CRAIG-MR methods using (5.17) and (5.50), with the difference that α_1 does not appear explicitly.

In order for the discussion of this section to apply to all four methods, we introduce the notation

$$\hat{\mathbf{T}}_k := \hat{\mathbf{U}}_k^T\hat{\mathbf{U}}_k,$$

where the value of $\hat{\mathbf{U}}_k$ is given in Table 5.1 for each method based on reduced equations. Note that $\hat{\mathbf{U}}_k$ is upper bidiagonal and that $\hat{\mathbf{T}}_k$ is tridiagonal, symmetric, and positive definite. From Theorems 5.2, 5.4, 5.6, and 5.8, the error estimates are related to the approximation of the leading entry $\hat{\mathbf{T}}_{11}^{-1}$ by $(\hat{\mathbf{T}}_k^{-1})_{11}$.

The vector $\mathbf{z}$ of entries ζ_j defines different quantities depending on the algorithm choice. In G-LSQR and G-LSMR,

$$\|\mathbf{z}\|_2^2 = \|\mathbf{A}^T\mathbf{M}^{-1}\mathbf{b}\|^2_{\mathbf{N}^{-1}}\,\hat{\mathbf{T}}_{11}^{-1} \quad \text{and} \quad \sum_{j=k+1}^{n}\zeta_j^2 = \|\mathbf{A}^T\mathbf{M}^{-1}\mathbf{b}\|^2_{\mathbf{N}^{-1}}\left(\hat{\mathbf{T}}_{11}^{-1} - (\hat{\mathbf{T}}_k^{-1})_{11}\right), \tag{5.56}$$

while in G-CRAIG and G-CRAIG-MR,

$$\|\mathbf{z}\|_2^2 = \|\mathbf{b}\|^2_{\mathbf{M}^{-1}}\,\hat{\mathbf{T}}_{11}^{-1} \quad \text{and} \quad \sum_{j=k+1}^{m}\zeta_j^2 = \|\mathbf{b}\|^2_{\mathbf{M}^{-1}}\left(\hat{\mathbf{T}}_{11}^{-1} - (\hat{\mathbf{T}}_k^{-1})_{11}\right), \tag{5.57}$$

where $\hat{\mathbf{T}}_k$ is the $k \times k$ principal submatrix of $\hat{\mathbf{T}}$.

Let $0 < \lambda_1 \le \cdots \le \lambda_p$ be the eigenvalues of the matrix $\mathbf{W}$ defining the energy norm in the method chosen, where p is either m or n. As explained in Theorems 5.2, 5.4, 5.6, and 5.8 and summarized in Table 5.1, $\mathbf{W}$ is also the matrix of the system being solved. Therefore, $\mathbf{W}$ and $\hat{\mathbf{T}}$ have the same eigenvalues.

Let $\hat{\mathbf{T}} = \mathbf{Q}\Lambda\mathbf{Q}^T$ be the eigendecomposition of $\hat{\mathbf{T}}$, where $\mathbf{Q}$ is orthogonal and $\Lambda = \operatorname{diag}(\lambda_1, \ldots, \lambda_p)$. The squared energy norm of the solution may then be expressed as

$$\|\mathbf{z}\|_2^2 = \gamma^2\mathbf{e}_1^T\hat{\mathbf{T}}^{-1}\mathbf{e}_1 = \gamma^2\mu^T\Lambda^{-1}\mu = \gamma^2\sum_{i=1}^{p}\lambda_i^{-1}\mu_i^2, \tag{5.58}$$

Table 5.1. *Factor $\hat{\mathbf{U}}_k$ of the tridiagonal and matrix $\mathbf{W}$ of the system solved by each method based on a reduced system. The same matrix $\mathbf{W}$ defines the energy norm for the method.*

Method	$\hat{\mathbf{U}}_k$	$\mathbf{W}$	ς_k	ν_k	Theorems
G-LSQR	$\mathbf{R}_k$	$\mathbf{A}^T\mathbf{M}^{-1}\mathbf{A}+\mathbf{N}$	ρ_k	θ_k	5.1, 5.2
G-CRAIG	$\hat{\mathbf{B}}_k^T$	$\mathbf{A}\mathbf{N}^{-1}\mathbf{A}^T+\mathbf{M}$	$\hat{\alpha}_k$	$\hat{\beta}_k$	5.3, 5.4
G-LSMR	$\mathbf{R}_k$	$(\mathbf{A}^T\mathbf{M}^{-1}\mathbf{A}+\mathbf{N})\mathbf{N}^{-1}(\mathbf{A}^T\mathbf{M}^{-1}\mathbf{A}+\mathbf{N})$	ρ_k	θ_k	5.5, 5.6
G-CRAIG-MR	$\hat{\mathbf{B}}_k^T$	$(\mathbf{A}\mathbf{N}^{-1}\mathbf{A}^T+\mathbf{M})\mathbf{M}^{-1}(\mathbf{A}\mathbf{N}^{-1}\mathbf{A}^T+\mathbf{M})$	$\hat{\alpha}_k$	$\hat{\beta}_k$	5.7, 5.8

where $\gamma > 0$ is either $\|\mathbf{A}^T\mathbf{M}^{-1}\mathbf{b}\|_{\mathbf{N}^{-1}}$ or $\|\mathbf{b}\|_{\mathbf{M}^{-1}}$ and $\mu := \mathbf{Q}^T\mathbf{e}_1 = (\mu_1,\ldots,\mu_p)$. Note that the components of μ are the first components of the normalized eigenvectors of $\hat{\mathbf{T}}$. Similarly, the squared energy norm of the kth approximation may be written as

$$\|\mathbf{z}_k\|_2^2 = \gamma^2 \sum_{i=1}^{k} \lambda_i^{-1}\mu_i^2,$$

so that the squared energy norm of the error is given by

$$\|\mathbf{z}-\mathbf{z}_k\|_2^2 = \gamma^2 \sum_{i=k+1}^{p} \lambda_i^{-1}\mu_i^2.$$

As earlier, we may choose to terminate the iterations as soon as

$$\sum_{i=k-d+1}^{k} \lambda_i^{-1}\mu_i^2 \le \tau^2 \sum_{i=1}^{k} \lambda_i^{-1}\mu_i^2 \tag{5.59}$$

for a given window size $d \in \mathbb{N}_0$ and tolerance $\tau^2 > 0$. Note that this relative stopping test does not depend on γ.

Following the exposition of Golub and Meurant (1997, 2010), the energy norm of $\mathbf{z}$ can be interpreted as the approximation by a Gauss quadrature of the Riemann–Stieltjes integral

$$\int_{\lambda_1}^{\lambda_p} \frac{1}{\lambda}\,\mathrm{d}\mu(\lambda), \tag{5.60}$$

where the measure μ is the nondecreasing step function

$$\mu(\lambda) = \begin{cases} 0 & \text{if } \lambda < \lambda_1, \\ \sum_{i=1}^{k} \mu_i^2 & \text{if } \lambda_k \le \lambda < \lambda_{k+1}, \\ \sum_{i=1}^{p} \mu_i^2 & \text{if } \lambda \ge \lambda_p. \end{cases}$$

Comparing the last sum in (5.58) with (5.60), we see that in this Gauss approximation, the nodes are given by the eigenvalues of $\hat{\mathbf{T}}$, while the weights are the squared first components of the normalized eigenvectors of $\hat{\mathbf{T}}$. Following this interpretation, the errors (5.56) and (5.57) can be viewed as the remainder of the approximation of (5.60) by this Gauss quadrature. In practice, there is no need to compute explicitly the eigenvalues λ_i or the weights μ_i—we simply accumulate the terms of the quadrature by computing recursively $\mathbf{e}_1^T\hat{\mathbf{T}}_k^{-1}\mathbf{e}_1$ for all k, stopping when new terms no longer add significantly to the overall sum.

The interesting feature of the above interpretation, as noted by Golub and Meurant (2010), is that because the sign of the remainder can be known in advance, the quadrature approximation will yield either an upper or a lower bound on the energy norm. This is due to the sign of the derivatives of $\lambda \mapsto 1/\lambda$ being known in advance. A pure Gauss approximation can be shown to yield a lower bound. But other quadratures are possible, such as the Gauss–Radau quadrature, in which one node is fixed, or the Gauss–Lobatto quadrature, in which two nodes are fixed. Fixing nodes amounts to augmenting $\hat{\mathbf{T}}_k$ so as to give it one or two prescribed eigenvalues. It can be shown that the Gauss–Radau rule yields an upper bound if $a \le \lambda_1$ is the fixed node or a lower bound if $b \ge \lambda_p$ is the fixed node, while the Gauss–Lobatto rule, in which both $a \le \lambda_1$ and $b \ge \lambda_p$ are fixed nodes, yields an upper bound. Note that the measure $\mu(\lambda)$ ensures that

$$\int_a^b \frac{1}{\lambda}\,\mathrm{d}\mu(\lambda) = \int_{\lambda_1}^{\lambda_p} \frac{1}{\lambda}\,\mathrm{d}\mu(\lambda)$$

for all such values of a and b.

We denote by ς_j the diagonal entries of $\hat{\mathbf{U}}$ and by ν_j the subdiagonal entries, so that

$$\hat{\mathbf{T}}_k = \begin{bmatrix} \varsigma_1 & & & \\ \nu_2 & \varsigma_2 & & \\ & \ddots & \ddots & \\ & & \nu_k & \varsigma_k \end{bmatrix} \begin{bmatrix} \varsigma_1 & \nu_2 & & \\ & \varsigma_2 & \ddots & \\ & & \ddots & \nu_k \\ & & & \varsigma_k \end{bmatrix} = \begin{bmatrix} \varsigma_1^2 & \varsigma_1\nu_2 & & \\ \varsigma_1\nu_2 & \varsigma_2^2+\nu_2^2 & \ddots & \\ & \ddots & \ddots & \varsigma_{k-1}\nu_k \\ & & \varsigma_{k-1}\nu_k & \varsigma_k^2+\nu_k^2 \end{bmatrix}.$$

Note that in all four methods, $\hat{\mathbf{T}}_k$ has the form $\mathbf{E}_k^{\mathsf{T}}\mathbf{E}_k + \mathbf{I}_k$ or $\mathbf{B}_k\mathbf{B}_k^{\mathsf{T}} + \mathbf{I}_k$. In both cases, $\lambda_1 \ge 1$. Let $0 < a < 1$ be a lower bound on all the eigenvalues of $\hat{\mathbf{T}}$. We now follow Golub and Meurant (2010) and describe how to implement the Gauss–Radau rule with a fixed node at a, thereby obtaining an upper bound on the error. When advancing from $\hat{\mathbf{T}}_k$ to $\hat{\mathbf{T}}_{k+1}$, we compute the new off-diagonal element $\varsigma_k\nu_{k+1}$ and modify the new diagonal element so that $\hat{\mathbf{T}}_{k+1}$ has an eigenvalue equal to a. The key element to setting the appropriate value on the diagonal resides in the relationship between the Lanczos process and the family of normalized polynomials $p_i(\lambda)$ that are orthogonal with respect to the measure μ, i.e.,

$$\int_{\lambda_1}^{\lambda_p} p_i(\lambda)p_j(\lambda)\mathrm{d}\mu(\lambda) = \delta_{ij},$$

where $\delta_{ij} = 1$ if $i = j$ and zero otherwise is the Kronecker symbol. Like all orthogonal polynomials, these satisfy a three-term recurrence relationship that is none other than, in matrix form,

$$\lambda \mathbf{p}_{k+1}(\lambda) = \hat{\mathbf{T}}_{k+1}\mathbf{p}_{k+1}(\lambda) + \varsigma_{k+1}\nu_{k+2}p_{k+1}(\lambda)\mathbf{e}_{k+1}, \tag{5.61}$$

where $\mathbf{p}_{k+1}(\lambda) = (p_0(\lambda), \dots, p_k(\lambda))$. From this relation, it becomes apparent that $\lambda = a$ is an eigenvalue of $\hat{\mathbf{T}}_{k+1}$ if and only if $p_{k+1}(a) = 0$. In this case, the last component equation of (5.61) reads

$$a p_k(a) = \varsigma_k\nu_{k+1}p_{k-1}(a) + (\varsigma_{k+1}^2 + \nu_{k+1}^2)p_k(a).$$

We now modify the $(\varsigma_{k+1}^2 + \nu_{k+1}^2)$ on the diagonal and replace it with the value that ensures satisfaction of this last identity, i.e.,

$$\omega_{k+1} := a - \varsigma_k \nu_{k+1} \frac{p_{k-1}(a)}{p_k(a)}. \tag{5.62}$$

The modified tridiagonal may be written as

$$\widetilde{\mathbf{T}}_{k+1} = \begin{bmatrix} \hat{\mathbf{T}}_k & \varsigma_k \nu_{k+1} \mathbf{e}_k \\ \varsigma_k \nu_{k+1} \mathbf{e}_k^\mathsf{T} & \omega_{k+1} \end{bmatrix}. \tag{5.63}$$

By construction, the smallest eigenvalue of $\widetilde{\mathbf{T}}_{k+1}$ is precisely a. The next difficulty is that the polynomials $p_i(\lambda)$ are not directly accessible. Fortunately, it is possible to evaluate ω_{k+1} by extracting the kth component δ_k of the solution $\boldsymbol{\delta}_k$ of a symmetric, positive-definite, and tridiagonal system. To see this, note that (5.61) at iteration k evaluated at $\lambda = a$ may be written as

$$(\hat{\mathbf{T}}_k - a\mathbf{I}_k)\boldsymbol{\delta}_k = \varsigma_k^2 \nu_{k+1}^2 \mathbf{e}_k, \tag{5.64}$$

where

$$\boldsymbol{\delta}_k = (\delta_1, \ldots, \delta_k) := -\frac{\varsigma_k \nu_{k+1}}{p_k(a)} \mathbf{p}_k(a) = -\varsigma_k \nu_{k+1} \left(\frac{p_0(a)}{p_k(a)}, \ldots, \frac{p_{k-1}(a)}{p_k(a)} \right).$$

Therefore, (5.62) may equivalently be written as

$$\omega_{k+1} = a + \delta_k.$$

Analogously to Arioli (2013) and Golub and Meurant (2010), we can recursively compute δ_k and ω_{k+1} by using the Cholesky decomposition for the system (5.64). Let

$$\hat{\mathbf{T}}_k - a\mathbf{I}_k = \begin{bmatrix} \varsigma_1^2 - a & \varsigma_1 \nu_2 & & \\ \varsigma_1 \nu_2 & \varsigma_2^2 + \nu_2^2 - a & \ddots & \\ & \ddots & \ddots & \varsigma_{k-1} \nu_k \\ & & \varsigma_{k-1} \nu_k & \varsigma_k^2 + \nu_k^2 - a \end{bmatrix}$$
$$= \begin{bmatrix} \ell_1 & & & \\ c_2 & \ell_2 & & \\ & \ddots & \ddots & \\ & & c_k & \ell_k \end{bmatrix} \begin{bmatrix} \ell_1 & c_2 & & \\ & \ell_2 & \ddots & \\ & & \ddots & c_k \\ & & & \ell_k \end{bmatrix}.$$

The Cholesky factors above are given by the recurrence relations

$$\ell_1 = \sqrt{\varsigma_1^2 - a}, \quad c_j = \varsigma_{j-1} \nu_j / \ell_{j-1}, \quad \ell_j = \sqrt{\varsigma_j^2 + \nu_j^2 - a - c_j^2}, \quad j = 2, 3, \ldots.$$

We may now compute δ_k by first solving

$$\begin{bmatrix} \ell_1 & & & \\ c_2 & \ell_2 & & \\ & \ddots & \ddots & \\ & & c_k & \ell_k \end{bmatrix} \begin{bmatrix} \pi_1 \\ \pi_2 \\ \vdots \\ \pi_k \end{bmatrix} = \begin{bmatrix} 0 \\ 0 \\ \vdots \\ \varsigma_k^2 \nu_{k+1}^2 \end{bmatrix},$$

i.e., $\pi_k = (\pi_1, \dots, \pi_k) = \varsigma_k^2 \nu_{k+1}^2 / \ell_k \, \mathbf{e}_k$, and next extracting the last component of the solution to

$$\begin{bmatrix} \ell_1 & c_2 & & \\ & \ell_2 & \ddots & \\ & & \ddots & c_k \\ & & & \ell_k \end{bmatrix} \begin{bmatrix} \delta_1 \\ \delta_2 \\ \vdots \\ \delta_k \end{bmatrix} = \begin{bmatrix} 0 \\ 0 \\ \vdots \\ \varsigma_k^2 \nu_{k+1}^2 / \ell_k \end{bmatrix},$$

i.e.,

$$\delta_k := \frac{\varsigma_k^2 \nu_{k+1}^2}{\ell_k^2} = \frac{\varsigma_k^2 \nu_{k+1}^2}{\varsigma_k^2 + \nu_k^2 - a - c_k^2},$$

with the special case that $\nu_k = c_k = 0$ when $k = 1$.

It remains to accumulate the terms of the quadrature. Having chosen a Gauss–Radau approximation to (5.60), we are interested in computing $\mathbf{e}_1^{\mathsf{T}} \widetilde{\mathbf{T}}_{k+1}^{-1} \mathbf{e}_1$ with $\widetilde{\mathbf{T}}_{k+1}$ defined as in (5.63). The Cholesky factorization $\widetilde{\mathbf{L}}_{k+1} \widetilde{\mathbf{L}}_{k+1}^{\mathsf{T}}$ of $\widetilde{\mathbf{T}}_{k+1}$ is given by

$$\widetilde{\mathbf{T}}_{k+1} = \begin{bmatrix} \hat{\mathbf{T}}_k & \varsigma_k \nu_{k+1} \mathbf{e}_k \\ \varsigma_k \nu_{k+1} \mathbf{e}_k^{\mathsf{T}} & \omega_{k+1} \end{bmatrix} = \begin{bmatrix} \hat{\mathbf{U}}_k^{\mathsf{T}} & \\ \nu_{k+1} \mathbf{e}_k^{\mathsf{T}} & u_{k+1} \end{bmatrix} \begin{bmatrix} \hat{\mathbf{U}}_k & \nu_{k+1} \mathbf{e}_k \\ & u_{k+1} \end{bmatrix} = \widetilde{\mathbf{L}}_{k+1} \widetilde{\mathbf{L}}_{k+1}^{\mathsf{T}},$$

where $u_{k+1} = \sqrt{\omega_{k+1} - \nu_{k+1}^2}$. Then $\mathbf{e}_1^{\mathsf{T}} \widetilde{\mathbf{T}}_{k+1}^{-1} \mathbf{e}_1 = \|\widetilde{\mathbf{L}}_{k+1}^{-1} \mathbf{e}_1\|_2^2$, and this last squared norm is accumulated into the variable Ξ^2 using the procedure

$$\begin{aligned} \chi_1 &= 1/\varsigma_1, & \Xi^2 &\leftarrow \chi_1^2, \\ \chi_{j+1} &= -\nu_{j+1} \chi_j / \varsigma_{j+1}, & \Xi^2 &\leftarrow \Xi^2 + \chi_{j+1}^2, \quad j = 1, 2, \dots, k-1, \\ \chi_{k+1} &= -\nu_{k+1} \chi_k / u_{k+1}, & \Xi^2 &\leftarrow \Xi^2 + \chi_{k+1}^2. \end{aligned}$$

Observe that, by definition, $\mathbf{z}_k = \gamma \hat{\mathbf{U}}_k^{-\mathsf{T}} \mathbf{e}_1$, and therefore the first k components of $\chi_k = (\chi_1, \dots, \chi_k)$ are precisely equal to $\gamma^{-1} \mathbf{z}_k$. Since the main loop of the algorithm already computes $\mathbf{z}_k$, we need only compute the last $\chi_{k+1} = -\nu_{k+1} \zeta_k / (\gamma u_{k+1})$. Since the γ in the denominator appears both in the components of χ_k and in Ξ, we simplify it in both places.

Finally, we obtain the implementation of the Gauss–Radau convergence test described in Algorithm 5.5. This algorithm should be interleaved with one of Algorithm 5.1, 5.2, 5.3, or 5.4, which we indicate with comments in lines 1 and 10.

Theorem 6.4 of Golub and Meurant (2010) ensures that Algorithm 5.5 computes an upper bound on the direct error. In typical situations, the major inconvenience of the Gauss–Radau approach is the need for an accurate estimate of the smallest eigenvalue of $\hat{\mathbf{T}}$, which can be very difficult in general. It is remarkable that in our case, the nature of $\hat{\mathbf{T}}$ guarantees that any value $0 < a < 1$ is such an estimate and produces an upper bound on the error.

The choice of the delay d is driven by the application and the values of $\mathbf{M}$ and $\mathbf{N}$. If $\mathbf{M}$ and $\mathbf{N}$ can be chosen such that the elliptic singular values of $\mathbf{A}$ become bounded in an interval independent of n and m, the delay parameter d can be quite small. We expect this to be the case in certain fluid flow problems, such as Stokes and stabilized Stokes, for $\mathbf{M}$ and $\mathbf{N}$ are then spectrally equivalent to operators defining appropriate norms in the relevant function spaces.

Algorithm 5.5 Gauss–Radau Convergence Test

Require: $\hat{\mathbf{U}}_k$, $d \in \mathbb{N}_0$, $\tau \in (0,1)$, $a \in (0,1)$

// Generate ς_1 and ζ_1

Set $\nu_1 := 0$, $c_1 := 0$, $\chi_1 := \zeta_1$, $\Xi_1^2 := \chi_1^2$, $k := 1$ and converged := **false.**

while $k < k_{\max}$ **do**

$\quad \ell_k^2 = \varsigma_k^2 + \nu_k^2 - a - c_k^2, \quad \delta_k = \varsigma_k^2 \nu_{k+1}^2 / \ell_k^2, \quad c_{k+1}^2 = \nu_{k+1}^2 / \ell_k^2$

$\quad \omega_{k+1} = a + \delta_k, \quad u_{k+1} = \sqrt{\omega_{k+1} - \nu_{k+1}^2}$

$\quad \chi_{k+1} = -\nu_{k+1} \zeta_k / u_{k+1}, \quad \Xi_k^2 = \sum_{j=1}^{k} \zeta_j^2$

$\quad$ **if** $k \geq d$ **then**

$\quad\quad$ converged $= \left(\sum_{j=k-d+1}^{k} \zeta_j^2 + \chi_{k+1}^2 \leq \tau \Xi_k^2 \right)$

$\quad$ **end if**

// Compute ζ_{k+1}

end while

Chapter 6

Full-Space Iterative Methods

In this chapter, we investigate the relation between the methods of Chapter 5 and two well-known Lanczos-based methods applied directly to the SQD system (1.5) with appropriate right-hand side: the conjugate-gradient method and the minimum residual method.

6.1 ▪ Full-Space Lanczos Process: I

Upon pasting the relations (4.3) together, we have

$$\begin{bmatrix} \mathbf{M} & \mathbf{A} \\ \mathbf{A}^\mathsf{T} & -\mathbf{N} \end{bmatrix} \begin{bmatrix} \mathbf{U}_{k+1} & \\ & \mathbf{V}_k \end{bmatrix} = \begin{bmatrix} \mathbf{M} & \\ & \mathbf{N} \end{bmatrix} \begin{bmatrix} \mathbf{U}_{k+1} & \\ & \mathbf{V}_k \end{bmatrix} \begin{bmatrix} \mathbf{I}_{k+1} & \mathbf{E}_k \\ \mathbf{E}_k^\mathsf{T} & -\mathbf{I}_k \end{bmatrix} + \begin{bmatrix} 0 \\ \alpha_{k+1} \mathbf{N} \mathbf{v}_{k+1} \end{bmatrix} \mathbf{e}_{2k+1}^\mathsf{T}, \tag{6.1a}$$

$$\begin{bmatrix} \mathbf{M} & \mathbf{A} \\ \mathbf{A}^\mathsf{T} & -\mathbf{N} \end{bmatrix} \begin{bmatrix} \mathbf{U}_{k} & \\ & \mathbf{V}_k \end{bmatrix} = \begin{bmatrix} \mathbf{M} & \\ & \mathbf{N} \end{bmatrix} \begin{bmatrix} \mathbf{U}_{k} & \\ & \mathbf{V}_k \end{bmatrix} \begin{bmatrix} \mathbf{I}_{k} & \mathbf{B}_k \\ \mathbf{B}_k^\mathsf{T} & -\mathbf{I}_k \end{bmatrix} + \begin{bmatrix} \beta_{k+1} \mathbf{M} \mathbf{u}_{k+1} \\ 0 \end{bmatrix} \mathbf{e}_{2k}^\mathsf{T}. \tag{6.1b}$$

We claim that (6.1) describes a Lanczos process on the matrix of (1.5) using a metric defined by blkdiag($\mathbf{M}, \mathbf{N}$) and that the matrix generated after k steps is itself SQD. Indeed, using the permutation matrix $\mathbf{P} := \begin{bmatrix} \mathbf{e}_1 & \mathbf{e}_{k+2} & \mathbf{e}_2 & \mathbf{e}_{k+3} & \cdots & \mathbf{e}_k & \mathbf{e}_{2k+1} & \mathbf{e}_{k+1} \end{bmatrix}$, we have

$$\mathbf{P}^\mathsf{T} \begin{bmatrix} \mathbf{I}_{k+1} & \mathbf{E}_k \\ \mathbf{E}_k^\mathsf{T} & -\mathbf{I}_k \end{bmatrix} \mathbf{P} = \mathbf{T}_{2k+1} := \begin{bmatrix} 1 & \alpha_1 & & & & \\ \alpha_1 & -1 & \beta_1 & & & \\ & \beta_1 & 1 & \alpha_2 & & \\ & & \ddots & \ddots & \ddots & \\ & & & \alpha_k & -1 & \beta_{k+1} \\ & & & & \beta_{k+1} & 1 \end{bmatrix} \tag{6.2}$$

$$= \begin{bmatrix} \mathbf{T}_{2k} & \beta_{k+1} \mathbf{e}_{2k} \\ \beta_{k+1} \mathbf{e}_{2k}^\mathsf{T} & 1 \end{bmatrix}, \tag{6.3}$$

which is the tridiagonal matrix $\mathbf{T}_{2k+1}$ generated after $2k+1$ steps of the Lanczos process described above. The Lanczos vectors $\mathbf{s}_k$ generated by the above process have the form

$s_{2k+1} := (u_k, 0)$ and $s_{2k+2} := (0, v_k)$ for $k \geq 0$. Moreover, the permutation $\mathbf{P}$ restores the order in which those vectors are generated by the algorithm, i.e.,

$$\mathbf{P}^\mathsf{T} \begin{bmatrix} \mathbf{U}_{k+1} & \\ & \mathbf{V}_k \end{bmatrix} \mathbf{P} = \begin{bmatrix} s_1 & s_2 & \cdots & s_{2k} & s_{2k+1} \end{bmatrix}.$$

Thus, the generalized Golub–Kahan process applied to $\mathbf{A}$ with initial vector $\mathbf{b}$ and metrics defined by $\mathbf{M}$ and $\mathbf{N}$ is equivalent to the Lanczos process applied to

$$\begin{bmatrix} \mathbf{M} & \mathbf{A} \\ \mathbf{A}^\mathsf{T} & -\mathbf{N} \end{bmatrix}$$

with initial vector $(\mathbf{b}, 0)$ and metric defined by blkdiag$(\mathbf{M}, \mathbf{N})$.

6.2 ▪ Relation with the Direct Lanczos Method

According to Definition 1.1, $\mathbf{T}_{2k}$ and $\mathbf{T}_{2k+1}$ defined in (6.2) and (6.3) are symmetric and quasi-definite. They therefore possess the Cholesky-like factorizations without pivoting:

$$\mathbf{T}_{2k} = \mathbf{L}_{2k}\mathbf{D}_{2k}\mathbf{L}_{2k}^\mathsf{T},$$

and, from (6.2) and (6.3),

$$\begin{aligned} \mathbf{T}_{2k+1} &= \begin{bmatrix} \mathbf{L}_{2k}\mathbf{D}_{2k}\mathbf{L}_{2k}^\mathsf{T} & \beta_{k+1}\mathbf{e}_{2k} \\ \beta_{k+1}\mathbf{e}_{2k}^\mathsf{T} & 1 \end{bmatrix} \\ &= \begin{bmatrix} \mathbf{L}_{2k} & \\ \ell_{2k}\mathbf{e}_{2k}^\mathsf{T} & 1 \end{bmatrix} \begin{bmatrix} \mathbf{D}_{2k} & \\ & d_{2k+1} \end{bmatrix} \begin{bmatrix} \mathbf{L}_{2k}^\mathsf{T} & \ell_{2k}\mathbf{e}_{2k} \\ & 1 \end{bmatrix} \\ &= \mathbf{L}_{2k+1}\mathbf{D}_{2k+1}\mathbf{L}_{2k+1}^\mathsf{T}, \end{aligned}$$

where $d_{2k+1} > 0$. Similarly, $\mathbf{T}_{2k+2}$ is given by

$$\begin{bmatrix} \mathbf{L}_{2k+1} & \\ \ell_{2k+1}\mathbf{e}_{2k+1}^\mathsf{T} & 1 \end{bmatrix} \begin{bmatrix} \mathbf{D}_{2k+1} & \\ & d_{2k+2} \end{bmatrix} \begin{bmatrix} \mathbf{L}_{2k+1}^\mathsf{T} & \ell_{2k+1}\mathbf{e}_{2k+1} \\ & 1 \end{bmatrix} = \mathbf{L}_{2k+2}\mathbf{D}_{2k+2}\mathbf{L}_{2k+2}^\mathsf{T},$$

where $d_{2k+2} < 0$. The factors are computed using the recursions

$$d_1 = 1, \quad d_{j+1} = t_{j+1,j+1} - d_j\ell_j^2, \quad \ell_j = t_{j+1,j}/d_j, \quad j = 1, 2, \ldots.$$

The entries $t_{i,j}$ of $\mathbf{T}_k$ are given by (6.2); $t_{j,j} = (-1)^{j+1}$, $t_{2j,2j-1} = \alpha_j$, and $t_{2j+1,2j} = \beta_j$.

The direct Lanczos method, referred to as DLANCZOS by Saad (2003, Algorithm 6.17), is simply the Lanczos process in which the tridiagonal system with coefficient $\mathbf{T}_k$ and with right-hand side $\beta_1\mathbf{e}_1$ is solved at each iteration. Our approach is to compute the factors $\mathbf{L}_k$ and $\mathbf{D}_k$ of $\mathbf{T}_k$ and update them at each iteration. Systems involving $\mathbf{T}_k$ are then solved by way of the usual forward and backward substitutions. Note that each $\mathbf{L}_k$ is unit lower bidiagonal.

Let $k = 2i$ be an even iteration number. Consider (1.5) with right-hand side $(\mathbf{b}, 0)$ and an approximation in the kth Krylov subspace of the form $(\mathbf{x}_k, \mathbf{y}_k) = (\mathbf{U}_k\bar{\mathbf{x}}_k, \mathbf{V}_k\bar{\mathbf{y}}_k)$. Upon premultiplying (1.5) with blkdiag$(\mathbf{U}_k^\mathsf{T}, \mathbf{V}_k^\mathsf{T})$ and using (6.1b), the $\mathbf{M}$-orthogonality of the vectors $\{\mathbf{u}_k\}$ and the $\mathbf{N}$-orthogonality of the vectors $\{\mathbf{v}_k\}$, we obtain (5.13), which is a step of G-CRAIG and is precisely the subproblem solved at iteration k of

DLANCZOS. Observe that the choice $\bar{y}_k = B_k^T \bar{x}_k$ automatically satisfies the second block of the equations (5.13) when $\bar{x}_k$ solves (5.17). Consider now an approximation in the $(k+1)$st Krylov space of the form $(x_{k+1}, y_k) = (U_{k+1}\bar{x}_{k+1}, V_k \bar{y}_k)$. Proceeding as above, we obtain (5.3), which is a step of G-LSQR and is precisely the subproblem solved at iteration $k+1$ of DLANCZOS. In particular, $\bar{y}_k$ satisfies the normal equations (5.1), and $\bar{x}_{k+1}$ satisfies the first $(k+1)$ equations of (5.3).

This behavior is a consequence of the expressions (2.28) and (2.29) for the Krylov spaces, which alternate between

$$\mathcal{K}_i\left(\bar{D}, \begin{bmatrix} b \\ 0 \end{bmatrix}\right) = \begin{bmatrix} \mathcal{K}_i(\bar{D}_1, b) \\ 0 \end{bmatrix} \quad \text{and} \quad \bar{K}\mathcal{K}_i\left(\bar{D}, \begin{bmatrix} b \\ 0 \end{bmatrix}\right) = \begin{bmatrix} \mathcal{K}_i(\bar{D}_1, b) \\ \mathcal{K}_i(\bar{D}_2, A^T b) \end{bmatrix}.$$

The above shows that the DLANCZOS method does not break down when applied to the SQD system (1.5) with $f = b$ and $g = 0$. Moreover, the underlying Lanczos process generates SQD matrices at each iteration. In particular, in exact arithmetic, the conjugate-gradient method alternates between the minimization of the convex part of the quadratic form generated by the SQD matrix and the maximization of its concave part. In other words, with $e_{x,k} = x - x_k$ and $e_{y,k} = y - y_k$, the DLANCZOS method solves the problem

$$\begin{aligned} \underset{x}{\text{minimize}}\, \underset{y}{\text{maximize}} \; & \begin{bmatrix} e_{x,k}^T & e_{y,k}^T \end{bmatrix} \begin{bmatrix} M & A \\ A^T & -N \end{bmatrix} \begin{bmatrix} e_{x,k} \\ e_{y,k} \end{bmatrix} \\ \text{subject to } & (x, y) \in \mathcal{K}_k\left(H^{-1}K, \begin{bmatrix} M^{-1}b \\ 0 \end{bmatrix}\right). \end{aligned} \tag{6.4}$$

A consequence of the previous paragraphs and the stability analysis of the LDL^T factorization of SQD matrices due to Gill, Saunders, and Shinnerl (1996) is that there is no need for a symmetric indefinite factorization of T_k in the vein of Marcia (2008) for problem (1.5).

We have proved the following result.

Theorem 6.1 (Relation between DLANCZOS, LSQR, and CRAIG). *The* DLANCZOS *method on the SQD system* (1.5) *with right-hand side* $(b, 0)$ *is well defined and will not break down. At each iteration, the tridiagonal matrix generated by the Lanczos process is SQD. Every odd step is a generalized* LSQR *step. Every even step is a generalized* CRAIG *step.*

If we denote $z_k = (x_k, y_k)$, $x_k = U_k \bar{x}_k$, $y_k = V_k \bar{y}_k$, and $\bar{z}_k = (\bar{x}_k, \bar{y}_k)$, the tridiagonal system (5.13) may be written as $T_{2k}\bar{z}_{2k} = \beta_1 e_1$. Taking into account the factorization of T_k, we obtain

$$z_k = S_k \bar{z}_k = S_k L_k^{-T} D_k^{-1} L_k^{-1}(\beta_1 e_1) = W_k q_k,$$

where we defined

$$W_k := S_k L_k^{-T} D_k^{-1} \qquad \text{and} \qquad q_k := L_k^{-1}(\beta_1 e_1).$$

This is equivalent to the usual derivation of the DLANCZOS method based on the LU factorization of the Lanczos tridiagonal. However, our usage of the LDL^T factorization highlights the fact that each T_k is SQD. The components $(\gamma_1, \ldots, \gamma_k)$ of q_k are easily found by recursion:

$$\gamma_1 = \beta_1, \qquad \gamma_{j+1} = -\ell_j \gamma_j, \quad j = 1,2,\ldots.$$

Similarly, using the equivalent identity $\mathbf{L}_k \mathbf{D}_k \mathbf{W}_k^{\mathsf{T}} = \mathbf{S}_k^{\mathsf{T}}$, we find the rows of $\mathbf{W}_k^{\mathsf{T}}$, i.e., vectors $\mathbf{w}_1, \ldots, \mathbf{w}_k$, recursively based on the vectors $\mathbf{s}_k$:

$$\mathbf{w}_1 = \mathbf{s}_1/d_1, \qquad \mathbf{w}_{j+1} = (\mathbf{s}_{j+1} - \ell_j d_j \mathbf{w}_j)/d_{j+1}, \quad j = 1,2,\ldots.$$

Knowledge of the vectors $\mathbf{w}_k$ leads to an efficient update of $\mathbf{z}_k$, bypassing the computation of $\bar{\mathbf{z}}_k$ altogether:

$$\mathbf{z}_k = \mathbf{W}_k \mathbf{q}_k = \mathbf{z}_{k-1} + \gamma_k \mathbf{w}_k.$$

As we show below, the matrix $\mathbf{W}_k$ forms a partial factor of $\mathbf{K}$, as defined in (2.11). This is a form of orthonormality of the vectors $\mathbf{w}_k$ in spite of the fact that $\mathbf{K}$ is indefinite. Indeed, the $\mathrm{LDL}^{\mathsf{T}}$ factorization of $\mathbf{T}_k$ can be arranged so that $\mathbf{L}_k$ is lower bidiagonal but not with unit diagonal, and $\mathbf{D}_k$ has ± 1 on its diagonal. With this alternative factorization, we have the updates

$$\gamma_1 = \beta_1/\ell_{1,1}, \qquad \gamma_{j+1} = -\ell_{j+1,j}\gamma_j/\ell_{j+1,j+1}, \quad j = 1,2,\ldots, \tag{6.5}$$

and

$$\mathbf{w}_1 = \mathbf{s}_1/\ell_{1,1}, \qquad \mathbf{w}_{j+1} = (\mathbf{s}_{j+1} - \ell_{j+1,j}\mathbf{w}_j)/\ell_{j+1,j+1}, \quad j = 1,2,\ldots. \tag{6.6}$$

Thus, without loss of generality, we may understand $\mathbf{D}_k$ as having diagonal elements $(-1)^{j+1}$.

Theorem 6.2 (Solution and Error Measures in DLANCZOS). *Let $\mathbf{W}_k$ be defined as above and let $\mathbf{T}_k = \mathbf{L}_k \mathbf{D}_k \mathbf{L}_k^{\mathsf{T}}$, where $\mathbf{L}_k$ is lower bidiagonal and $\mathbf{D}_k$ is diagonal. Then, for $k = 1, \ldots, n+m$, we have the partial factorization*

$$\mathbf{W}_k^{\mathsf{T}} \mathbf{K} \mathbf{W}_k = \mathbf{D}_k^{-1}. \tag{6.7}$$

The DLANCZOS *iterates satisfy*

$$\mathbf{z}_k = \sum_{j=1}^{k} \gamma_j \mathbf{w}_j \tag{6.8}$$

and

$$\mathbf{z}_k^{\mathsf{T}} \mathbf{K} \mathbf{z}_k = \sum_{j=1}^{k} \frac{\gamma_j^2}{d_j} = \sum_{j=1}^{\lceil k/2 \rceil} \frac{\gamma_{2j-1}^2}{d_{2j-1}} - \sum_{j=1}^{\lfloor k/2 \rfloor} \frac{\gamma_{2j}^2}{(-d_{2j})}, \tag{6.9}$$

as well as the error identity

$$(\mathbf{z} - \mathbf{z}_k)^{\mathsf{T}} \mathbf{K} (\mathbf{z} - \mathbf{z}_k) = \sum_{j=k+1}^{n+m} \frac{\gamma_j^2}{d_j} = \sum_{j=k+1}^{\lceil \frac{n+m}{2} \rceil} \frac{\gamma_{2j-1}^2}{d_{2j-1}} - \sum_{j=k+1}^{\lfloor \frac{n+m}{2} \rfloor} \frac{\gamma_{2j}^2}{(-d_{2j})}, \tag{6.10}$$

where $\mathbf{z}$ is an exact solution of (1.5) *with $\mathbf{f} = \mathbf{b}$ and $\mathbf{g} = 0$.*

Proof. It suffices to note that, after applying the permutation $\mathbf{P}$, we have from (6.1b)

$$\mathbf{S}_k^\top \mathbf{K} \mathbf{S}_k = \mathbf{S}_k^\top \mathbf{H} \mathbf{S}_k \mathbf{T}_k = \mathbf{L}_k^\top \mathbf{D}_k \mathbf{L}_k,$$

where $\mathbf{H}$ is defined in (2.12) and where we used the fact that the vectors $\mathbf{s}_k$ are orthonormal in the metric defined by $\mathbf{H}$. Introducing now the definition of $\mathbf{W}_k$, we obtain

$$\mathbf{W}_k^\top \mathbf{K} \mathbf{W}_k = \mathbf{D}_k^{-1} \mathbf{L}_k^{-1} \mathbf{S}_k^\top \mathbf{K} \mathbf{S}_k \mathbf{L}_k^{-\top} \mathbf{D}_k^{-1} = \mathbf{D}_k^{-1}.$$

The rest of the proof follows directly, as in previous sections. □

It is important to note that $\mathbf{K}$, being indefinite, does not define a norm. It does, however, define a distance, and this is why we do not use the norm notation in (6.9) and (6.10). Indeed, $d_{2k} < 0$, and $d_{2k+1} > 0$. We use the nomenclature of Gohberg, Lancaster, and Rodman (2005, Chapter 2). The vectors $\mathbf{w}_k$ are orthogonal in the metric $\mathbf{K}$ and orthonormal if the factorization of $\mathbf{T}_k$ is arranged so that

$$\mathbf{D}_{2k+1} = \mathbf{P}^\top \begin{bmatrix} \mathbf{I}_{k+1} & \\ & -\mathbf{I}_k \end{bmatrix} \mathbf{P}.$$

For these reasons, we emphasized the negative terms in (6.9) and (6.10) by separating odd and even indices. The sum over odd indices corresponds to G-CRAIG iterations where $d_{2j-1} > 0$ and where the primal, minimum-norm problem is solved. The sum over even indices corresponds to G-LSQR iterations where the dual, negative least-squares problem is solved.

The conjugate-gradient method is simply a reformulation of DLANCZOS, in which the LU factorization of $\mathbf{T}_k$ is computed instead. Therefore, the conclusions above also apply to CG with appropriate redefinitions of $\mathbf{L}_k$ and $\mathbf{D}_k$. Interestingly, even in the present indefinite context, the conjugate-gradient algorithm continues to perform its well-known minimization of the error in the energy "norm" with the difference that it alternates between minimization steps in one problem and maximization steps in the dual problem viewed as a maximization problem. In particular, it is possible to define a corresponding stopping test based on the direct error in "energy norm." Select an integer $d > 0$ and a threshold $\tau > 0$. We may terminate the DLANCZOS iterations as soon as the partial sums (6.10) computed only over the past d iterations stabilize. More precisely, we terminate the iterations when

$$\sum_{j=k+1}^{\lceil \frac{k+d+1}{2} \rceil} \frac{\gamma_{2j-1}^2}{d_{2j-1}} < \tau^2 \sum_{j=1}^{\lceil \frac{k+d+1}{2} \rceil} \frac{\gamma_{2j-1}^2}{d_{2j-1}} \quad \text{and} \quad \sum_{j=k+1}^{\lfloor \frac{k+d+1}{2} \rfloor} \frac{\gamma_{2j}^2}{(-d_{2j})} < \tau^2 \sum_{j=1}^{\lfloor \frac{k+d+1}{2} \rfloor} \frac{\gamma_{2j}^2}{(-d_{2j})}.$$

In exact arithmetic, it is equivalent to stopping as soon as

$$\sum_{\substack{j=k+1 \\ d_j>0}}^{k+d+1} \frac{\gamma_j^2}{d_j} < \tau^2 \sum_{\substack{j=1 \\ d_j>0}}^{k+d+1} \frac{\gamma_j^2}{d_j} \quad \text{and} \quad \sum_{\substack{j=k+1 \\ d_j<0}}^{k+d+1} \frac{\gamma_j^2}{(-d_j)} < \tau^2 \sum_{\substack{j=1 \\ d_j<0}}^{k+d+1} \frac{\gamma_j^2}{(-d_j)}. \tag{6.11}$$

Alternatively, it is also possible to stop as soon as

$$\sum_{j=k+1}^{k+d+1} \frac{\gamma_j^2}{|d_j|} < \tau^2 \sum_{j=1}^{k+d+1} \frac{\gamma_j^2}{|d_j|}.$$

We stress that the error (6.10) measured in the metric $\mathbf{K}$ can be either positive or negative. Although we have not formally established this fact, in practice, its sign typically alternates, as does the sign of the pivots d_j, and exhibits an oscillatory behavior. It approaches zero in absolute value as k approaches $n+m$. This is illustrated in the numerical experiments of Section 7.3.

6.3 ▪ Full-Space Lanczos Process: II

The two methods G-LSMR and G-CRAIG-MR of Sections 5.5 and 5.7 turn out to combine to become equivalent to an appropriately preconditioned MINRES on (1.5). This result parallels Theorem 6.1 and the combination of G-LSQR and G-CRAIG to form G-CG. Upon pasting the Lanczos processes (5.38) and (5.48) together, we obtain

$$\begin{aligned}
&\begin{bmatrix} \mathbf{M}+\mathbf{A}\mathbf{N}^{-1}\mathbf{A}^T & \\ & \mathbf{N}+\mathbf{A}^T\mathbf{M}^{-1}\mathbf{A} \end{bmatrix} \begin{bmatrix} \mathbf{U}_k & \\ & \mathbf{V}_k \end{bmatrix} \\
&\quad = \begin{bmatrix} \mathbf{M} & \\ & \mathbf{N} \end{bmatrix} \left(\begin{bmatrix} \mathbf{U}_k & \\ & \mathbf{V}_k \end{bmatrix} \begin{bmatrix} \mathbf{B}_k\mathbf{B}_k^T+\mathbf{I}_k & \\ & \mathbf{E}_k^T\mathbf{E}_k+\mathbf{I}_k \end{bmatrix} \right. \\
&\qquad \left. + \begin{bmatrix} \alpha_k\beta_{k+1}\mathbf{u}_{k+1}\mathbf{e}_k^T & \\ & \alpha_{k+1}\beta_{k+1}\mathbf{v}_{k+1}\mathbf{e}_k^T \end{bmatrix} \right).
\end{aligned} \tag{6.12}$$

As already noted in (2.22),

$$\begin{bmatrix} \mathbf{M} & \mathbf{A} \\ \mathbf{A}^T & -\mathbf{N} \end{bmatrix} \begin{bmatrix} \mathbf{M}^{-1} & \\ & \mathbf{N}^{-1} \end{bmatrix} \begin{bmatrix} \mathbf{M} & \mathbf{A} \\ \mathbf{A}^T & -\mathbf{N} \end{bmatrix} = \begin{bmatrix} \mathbf{M}+\mathbf{A}\mathbf{N}^{-1}\mathbf{A}^T & \\ & \mathbf{N}+\mathbf{A}^T\mathbf{M}^{-1}\mathbf{A} \end{bmatrix}.$$

If we premultiply (6.12) by $\operatorname{blkdiag}(\mathbf{U}_k^T, \mathbf{V}_k^T)$ and use the $\mathbf{M}$-orthogonality of the vectors $\mathbf{u}_k$ and the $\mathbf{N}$-orthogonality of the vectors $\mathbf{v}_k$, we see that the Lanczos process may be equivalently rewritten as

$$\begin{aligned}
&\begin{bmatrix} \mathbf{U}_k & \\ & \mathbf{V}_k \end{bmatrix}^T \begin{bmatrix} \mathbf{M}+\mathbf{A}\mathbf{N}^{-1}\mathbf{A}^T & \\ & \mathbf{N}+\mathbf{A}^T\mathbf{M}^{-1}\mathbf{A} \end{bmatrix} \begin{bmatrix} \mathbf{U}_k & \\ & \mathbf{V}_k \end{bmatrix} \\
&\quad = \begin{bmatrix} \mathbf{B}_k\mathbf{B}_k^T+\mathbf{I}_k & \\ & \mathbf{E}_k^T\mathbf{E}_k+\mathbf{I}_k \end{bmatrix} = \begin{bmatrix} \mathbf{B}_k\mathbf{B}_k^T+\mathbf{I}_k & \\ & \mathbf{B}_k^T\mathbf{B}_k+\mathbf{I}_k \end{bmatrix} + \beta_{k+1}^2\mathbf{e}_{2k}\mathbf{e}_{2k}^T.
\end{aligned}$$

6.4 ▪ Relation with the Minimum Residual Method

In MINRES, (1.5) with $\mathbf{f}=\mathbf{b}$ and $\mathbf{g}=0$ is tackled directly using sequential approximations in a Lanczos subspace by iteratively minimizing the norm of the residual. Typically, the Euclidean norm is used, but other norms can be used via preconditioning. It turns out that the relevant Lanczos process in this case is precisely (6.1) and the kth Krylov subspace $\mathscr{K}_k$ is spanned by the vectors $(\mathbf{u}_k, 0)$ and $(0, \mathbf{v}_k)$. Let $(\mathbf{x}_k, \mathbf{y}_k)$ be our approximation of $(\mathbf{x}, \mathbf{y})$ in $\mathscr{K}_k$. The corresponding residual is given by

$$\begin{bmatrix} \mathbf{r}_k \\ \mathbf{s}_k \end{bmatrix} := \begin{bmatrix} \mathbf{b} \\ 0 \end{bmatrix} - \begin{bmatrix} \mathbf{M} & \mathbf{A} \\ \mathbf{A}^T & -\mathbf{N} \end{bmatrix} \begin{bmatrix} \mathbf{x}_k \\ \mathbf{y}_k \end{bmatrix} \tag{6.13}$$

$$= \begin{bmatrix} \beta_1\mathbf{M}\mathbf{u}_1 \\ 0 \end{bmatrix} - \begin{bmatrix} \mathbf{M} & \mathbf{A} \\ \mathbf{A}^T & -\mathbf{N} \end{bmatrix} \begin{bmatrix} \mathbf{U}_k & \\ & \mathbf{V}_k \end{bmatrix} \begin{bmatrix} \bar{\mathbf{x}}_k \\ \bar{\mathbf{y}}_k \end{bmatrix} \tag{6.14}$$

for some vector $(\bar{\mathbf{x}}_k, \bar{\mathbf{y}}_k)$. In this section, we show that the Lanczos process (6.1) determines an implementation of MINRES in which the residual (6.13) is minimized in the norm defined by blkdiag$(\mathbf{M}^{-1}, \mathbf{N}^{-1})$, which requires that $(\bar{\mathbf{x}}_k, \bar{\mathbf{y}}_k)$ satisfy the Ritz–Galerkin condition

$$\begin{bmatrix} \mathbf{U}_k & \\ & \mathbf{V}_k \end{bmatrix}^{\mathsf{T}} \begin{bmatrix} \mathbf{M} & \mathbf{A} \\ \mathbf{A}^{\mathsf{T}} & -\mathbf{N} \end{bmatrix} \begin{bmatrix} \mathbf{M}^{-1} & \\ & \mathbf{N}^{-1} \end{bmatrix} \begin{bmatrix} \mathbf{r}_k \\ \mathbf{s}_k \end{bmatrix} = \begin{bmatrix} 0 \\ 0 \end{bmatrix}. \tag{6.15}$$

We premultiply (6.1b) with

$$\begin{bmatrix} \mathbf{U}_k & \\ & \mathbf{V}_k \end{bmatrix}^{\mathsf{T}} \begin{bmatrix} \mathbf{M} & \mathbf{A} \\ \mathbf{A}^{\mathsf{T}} & -\mathbf{N} \end{bmatrix}^{\mathsf{T}} \begin{bmatrix} \mathbf{M}^{-1} & \\ & \mathbf{N}^{-1} \end{bmatrix}$$

and obtain a sum of two terms. The first term is

$$\begin{bmatrix} \mathbf{U}_k^{\mathsf{T}} & \\ & \mathbf{V}_k^{\mathsf{T}} \end{bmatrix} \begin{bmatrix} \mathbf{M} & \mathbf{A} \\ \mathbf{A}^{\mathsf{T}} & -\mathbf{N} \end{bmatrix} \begin{bmatrix} \mathbf{U}_k & \\ & \mathbf{V}_k \end{bmatrix} \begin{bmatrix} \mathbf{I}_k & \mathbf{B}_k \\ \mathbf{B}_k^{\mathsf{T}} & -\mathbf{I}_k \end{bmatrix}.$$

Using (6.1b) itself, we may rewrite this term as

$$\begin{aligned} &\begin{bmatrix} \mathbf{U}_k^{\mathsf{T}} & \\ & \mathbf{V}_k^{\mathsf{T}} \end{bmatrix} \begin{bmatrix} \mathbf{M} & \\ & \mathbf{N} \end{bmatrix} \begin{bmatrix} \mathbf{U}_k & \\ & \mathbf{V}_k \end{bmatrix} \begin{bmatrix} \mathbf{I}_k & \mathbf{B}_k \\ \mathbf{B}_k^{\mathsf{T}} & -\mathbf{I}_k \end{bmatrix}^2 \\ &\quad + \begin{bmatrix} \mathbf{U}_k^{\mathsf{T}} & \\ & \mathbf{V}_k^{\mathsf{T}} \end{bmatrix} \begin{bmatrix} \beta_{k+1}\mathbf{M}\mathbf{u}_{k+1} \\ 0 \end{bmatrix} \mathbf{e}_{2k}^{\mathsf{T}} \begin{bmatrix} \mathbf{I}_k & \mathbf{B}_k \\ \mathbf{B}_k^{\mathsf{T}} & -\mathbf{I}_k \end{bmatrix} = \begin{bmatrix} \mathbf{I}_k & \mathbf{B}_k \\ \mathbf{B}_k^{\mathsf{T}} & -\mathbf{I}_k \end{bmatrix}^2, \end{aligned}$$

where we used the $\mathbf{M}$-orthogonality of $\{\mathbf{v}_1, \dots, \mathbf{v}_k\}$ and the $\mathbf{N}$-orthogonality of $\{\mathbf{u}_1, \dots, \mathbf{u}_k\}$. The second term is

$$\begin{bmatrix} \mathbf{U}_k & \\ & \mathbf{V}_k \end{bmatrix}^{\mathsf{T}} \begin{bmatrix} \mathbf{M} & \mathbf{A} \\ \mathbf{A}^{\mathsf{T}} & -\mathbf{N} \end{bmatrix} \begin{bmatrix} \mathbf{M}^{-1} & \\ & \mathbf{N}^{-1} \end{bmatrix} \begin{bmatrix} \beta_{k+1}\mathbf{M}\mathbf{u}_{k+1} \\ 0 \end{bmatrix} \mathbf{e}_{2k}^{\mathsf{T}} = \beta_{k+1}^2 \mathbf{e}_{2k}\mathbf{e}_{2k}^{\mathsf{T}},$$

where we used (6.1b) and the orthogonality properties of the Lanczos vectors one more time, in the same way as for the first term. We have just shown that

$$\begin{aligned} &\begin{bmatrix} \mathbf{U}_k & \\ & \mathbf{V}_k \end{bmatrix}^{\mathsf{T}} \begin{bmatrix} \mathbf{M} & \mathbf{A} \\ \mathbf{A}^{\mathsf{T}} & -\mathbf{N} \end{bmatrix} \begin{bmatrix} \mathbf{M}^{-1} & \\ & \mathbf{N}^{-1} \end{bmatrix} \begin{bmatrix} \mathbf{M} & \mathbf{A} \\ \mathbf{A}^{\mathsf{T}} & -\mathbf{N} \end{bmatrix} \begin{bmatrix} \mathbf{U}_k & \\ & \mathbf{V}_k \end{bmatrix} \\ &\qquad = \begin{bmatrix} \mathbf{I}_k & \mathbf{B}_k \\ \mathbf{B}_k^{\mathsf{T}} & -\mathbf{I}_k \end{bmatrix}^2 + \beta_{k+1}^2 \mathbf{e}_{2k}\mathbf{e}_{2k}^{\mathsf{T}} \end{aligned} \tag{6.16}$$

is the underlying Lanczos process governing the minimum residual method. The first matrix on the right-hand side of this last equality is the square of a symmetric permutation of $\mathbf{T}_{2k}$. It is thus a symmetric permutation of the pentadiagonal matrix $\mathbf{T}_{2k}^2$.

On the other hand, because

$$\begin{bmatrix} \mathbf{U}_k & \\ & \mathbf{V}_k \end{bmatrix}^{\mathsf{T}} \begin{bmatrix} \mathbf{M} & \mathbf{A} \\ \mathbf{A}^{\mathsf{T}} & -\mathbf{N} \end{bmatrix}^{\mathsf{T}} \begin{bmatrix} \mathbf{M}^{-1} & \\ & \mathbf{N}^{-1} \end{bmatrix} \begin{bmatrix} \beta_1\mathbf{M}\mathbf{u}_1 \\ 0 \end{bmatrix} = \begin{bmatrix} \beta_1\mathbf{e}_1 \\ \alpha_1\beta_1\mathbf{e}_1 \end{bmatrix},$$

(6.15) amounts to the (psychologically) pentadiagonal system

$$\left(\begin{bmatrix} \mathbf{I}_k & \mathbf{B}_k \\ \mathbf{B}_k^{\mathsf{T}} & -\mathbf{I}_k \end{bmatrix}^2 + \beta_{k+1}^2 \mathbf{e}_{2k}\mathbf{e}_{2k}^{\mathsf{T}} \right) \begin{bmatrix} \bar{\mathbf{x}}_k \\ \bar{\mathbf{y}}_k \end{bmatrix} = \begin{bmatrix} \beta_1\mathbf{e}_1 \\ \alpha_1\beta_1\mathbf{e}_1 \end{bmatrix}. \tag{6.17}$$

System (6.17) is precisely the system solved by the standard MINRES at iteration k (Paige and Saunders, 1975, Equations (6.1) and (6.2)). It is also easy to verify that (6.17) represents the optimality conditions of the linear least-squares problem

$$\underset{\bar{\mathbf{x}},\bar{\mathbf{y}}}{\text{minimize}}\ \frac{1}{2}\left\|\begin{bmatrix}\mathbf{I}_{k+1} & \mathbf{E}_k \\ \mathbf{E}_k^\mathsf{T} & -\mathbf{I}_k\end{bmatrix}\begin{bmatrix}\bar{\mathbf{x}} \\ \bar{\mathbf{y}}\end{bmatrix} - \begin{bmatrix}\beta_1 \mathbf{e}_1 \\ 0\end{bmatrix}\right\|_2^2.$$

As a consequence, we obtain the generalized MINRES by replacing the standard Lanczos process used at each iteration of MINRES with (6.1). We have just proved the following theorem.

Theorem 6.3 (Relation between MINRES, LSMR, and CRAIG-MR). *The generalized* MINRES *iterates on* (1.5) *with right-hand side* $(\mathbf{b}, 0)$ *are the same as those generated by* MINRES *on* (1.5) *with preconditioner* $\text{blkdiag}(\mathbf{M}^{-1}, \mathbf{N}^{-1})$. *In addition, every even step is a* G-LSMR *step, and every odd step is a* G-CRAIG-MR *step.*

Proof. It suffices to note that

$$\begin{bmatrix}\mathbf{B}_k\mathbf{B}_k^\mathsf{T} + \mathbf{I}_k & \\ & \mathbf{B}_k^\mathsf{T}\mathbf{B}_k + \mathbf{I}_k\end{bmatrix} = \begin{bmatrix}\mathbf{I}_k & \mathbf{B}_k \\ \mathbf{B}_k^\mathsf{T} & -\mathbf{I}_k\end{bmatrix}^2$$

and that the Lanczos processes (6.12) and (6.16) are identical. This common Lanczos process implies that the G-MINRES method applied to (1.5) with right-hand side $(\mathbf{b}, 0)$ and preconditioner $\text{blkdiag}(\mathbf{M}^{-1}, \mathbf{N}^{-1})$ alternates between G-CRAIG-MR steps and G-LSMR steps. ☐

By definition of MINRES, the quantity $\|\mathbf{M}^{-\frac{1}{2}}\mathbf{r}_k\|_2^2 + \|\mathbf{N}^{-\frac{1}{2}}\mathbf{s}_k\|_2^2 = \|\mathbf{r}_k\|_{\mathbf{M}^{-1}}^2 + \|\mathbf{s}_k\|_{\mathbf{N}^{-1}}^2$ is nonincreasing with k. As we now show, MINRES also lends itself to an interpretation in terms of direct error, as opposed to residual, and a related stopping condition emerges. We begin with generalities on MINRES that lead to a new result indicating the appropriate norm in which MINRES measures direct errors. Because this result is general and always applies to MINRES, it is given in a separate section. We next specialize this result to the current SQD framework to provide an interpretation of MINRES as a decoupled combination of G-LSMR and G-CRAIG-MR.

6.4.1 ▪ Generalities on MINRES

Consider a generic linear system $\mathbf{Hx} = \mathbf{d}$, where $\mathbf{H} = \mathbf{H}^\mathsf{T}$. MINRES generates Lanczos vectors $\mathbf{s}_k$ and a symmetric tridiagonal matrix $\boldsymbol{\Omega}_k$ according to (4.1). The process is summarized by (4.2). Much as in the previous section, MINRES can be summarized with the identity

$$\mathbf{S}_k^\mathsf{T}\mathbf{H}^2\mathbf{S}_k = \boldsymbol{\Omega}_k^2 + \beta_{k+1}^2\mathbf{e}_k\mathbf{e}_k^\mathsf{T}, \tag{6.18}$$

where the columns of $\mathbf{S}_k$ are theoretically orthonormal. Paige and Saunders (1975) compute the LQ factorization of $\boldsymbol{\Omega}_k$ and show that

$$\boldsymbol{\Omega}_k^2 + \beta_{k+1}^2\mathbf{e}_k\mathbf{e}_k^\mathsf{T} = \mathbf{L}_k\mathbf{L}_k^\mathsf{T}, \tag{6.19}$$

where $\mathbf{L}_k$ is lower tridiagonal. The iterates are updated according to

$$\mathbf{x}_k = \mathbf{W}_k \mathbf{t}_k = \sum_{j=1}^{k} \tau_j \mathbf{w}_j = \mathbf{x}_{k-1} + \tau_k \mathbf{w}_k,$$

where $\mathbf{t}_k = (\tau_1, \dots, \tau_k)$ and $\mathbf{L}_k \mathbf{W}_k^\mathsf{T} := \mathbf{S}_k^\mathsf{T}$. The scalar τ_k is easily obtained by way of a recurrence at each iteration. We refer the reader to Paige and Saunders (1975, Section 6) for details. The above is sufficient to establish the following result.

Theorem 6.4 (Solution and Error Norm in MINRES). *Let $\mathbf{W}_k$ be defined as above. Then, for $k = 1, \dots, n$, we have the partial factorization*

$$\mathbf{W}_k^\mathsf{T} \mathbf{H}^2 \mathbf{W}_k = \mathbf{I}_k. \tag{6.20}$$

The MINRES *iterates satisfy*

$$\mathbf{x}_k = \sum_{j=1}^{k} \tau_j \mathbf{w}_j \tag{6.21}$$

and

$$\|\mathbf{x}_k\|_{\mathbf{H}^2}^2 = \sum_{j=1}^{k} \tau_j^2, \tag{6.22}$$

as well as the error identity

$$\|\mathbf{x} - \mathbf{x}_k\|_{\mathbf{H}^2}^2 = \sum_{j=k+1}^{n} \tau_j^2, \tag{6.23}$$

where $\mathbf{x}$ *is the solution of* $\mathbf{H}\mathbf{x} = \mathbf{d}$.

Proof. It suffices to note that

$$\mathbf{W}_k^\mathsf{T} \mathbf{H}^2 \mathbf{W}_k = \mathbf{L}_k^{-1} \mathbf{S}_k^\mathsf{T} \mathbf{H}^2 \mathbf{S}_k \mathbf{L}_k^{-\mathsf{T}} = \mathbf{L}_k^{-1} \mathbf{L}_k \mathbf{L}_k^\mathsf{T} \mathbf{L}_k^{-\mathsf{T}} = \mathbf{I}_k,$$

where we used (6.18) and (6.19). The rest of the proof is analogous to those of previous similar results. □

In the next section, we specialize Theorem 6.4 to the SQD context.

6.4.2 ▪ An error estimate for MINRES

In the SQD context, the generic identity (6.18) is paralleled by (6.16). The matrix multiplied left and right by the Lanczos vectors on the left-hand side of (6.16) is precisely the block-diagonal matrix $\mathbf{D}$ given by (2.21). MINRES performs an LQ factorization of

$$\begin{bmatrix} \mathbf{I}_k & \mathbf{B}_k \\ \mathbf{B}_k^\mathsf{T} & -\mathbf{I}_k \end{bmatrix}$$

in such a way that

$$\begin{bmatrix} \mathbf{U}_k & \\ & \mathbf{V}_k \end{bmatrix}^\mathsf{T} \begin{bmatrix} \mathbf{M} + \mathbf{A}\mathbf{N}^{-1}\mathbf{A}^\mathsf{T} & \\ & \mathbf{N} + \mathbf{A}^\mathsf{T}\mathbf{M}^{-1}\mathbf{A} \end{bmatrix} \begin{bmatrix} \mathbf{U}_k & \\ & \mathbf{V}_k \end{bmatrix} = \mathbf{L}_k \mathbf{L}_k^\mathsf{T},$$

where $\mathbf{L}_k$ is lower tridiagonal. The following result is a simple rewrite of Theorem 6.4.

Theorem 6.5 (Solution and Error Norm in MINRES on SQD Systems). *Let $\mathbf{W}_k$ be defined as in Theorem 6.4. Then, for $k = 1, \ldots, n+m$, we have the partial factorization*

$$\mathbf{W}_k^{\mathsf{T}} \mathbf{D} \mathbf{W}_k = \mathbf{I}_k, \tag{6.24}$$

where $\mathbf{D}$ is defined in (2.21). In addition, the MINRES iterates satisfy

$$\begin{bmatrix} \mathbf{x}_k \\ \mathbf{y}_k \end{bmatrix} = \sum_{j=1}^{k} \tau_j \mathbf{w}_j \tag{6.25}$$

and

$$\|(\mathbf{x}_k, \mathbf{y}_k)\|_{\mathbf{D}}^2 = \|\mathbf{x}_k\|_{\mathbf{M}+\mathbf{A}\mathbf{N}^{-1}\mathbf{A}^{\mathsf{T}}}^2 + \|\mathbf{y}_k\|_{\mathbf{N}+\mathbf{A}^{\mathsf{T}}\mathbf{M}^{-1}\mathbf{A}}^2 = \sum_{j=1}^{k} \tau_j^2, \tag{6.26}$$

as well as the error identity

$$\begin{aligned} \|(\mathbf{x}-\mathbf{x}_k, \mathbf{y}-\mathbf{y}_k)\|_{\mathbf{D}}^2 &= \|\mathbf{x}-\mathbf{x}_k\|_{\mathbf{M}+\mathbf{A}\mathbf{N}^{-1}\mathbf{A}^{\mathsf{T}}}^2 + \|\mathbf{y}-\mathbf{y}_k\|_{\mathbf{N}+\mathbf{A}^{\mathsf{T}}\mathbf{M}^{-1}\mathbf{A}}^2 \\ &= \sum_{j=k+1}^{n+m} \tau_j^2, \end{aligned} \tag{6.27}$$

where $(\mathbf{x}, \mathbf{y})$ is the solution of (1.5) with $\mathbf{f} = \mathbf{b}$ and $\mathbf{g} = 0$.

In particular, applying MINRES on (1.5) consists in applying G-LSMR in the variable $\mathbf{y}$ and G-CRAIG-MR in the variable $\mathbf{x}$ in a decoupled manner. This comes from the fact that $\mathbf{D}$ is block diagonal. It is then clear that, by contrast with CG, MINRES performs twice the work, as it applies both G-LSMR and G-CRAIG-MR and terminates when both have reached satisfactory accuracy.

Note that the energy norm given in (6.26) and (6.27) differs from that of Silvester and Simoncini (2011), who use the norm defined by the matrix $\mathbf{H}$ of (2.12) in the context of systems of PDEs. The two error norms are related in some cases, such as the simulation of Stokes flows by way of a mixed finite-element discretization. In this case, $\mathbf{N}$ is typically assumed to be spectrally equivalent to the appropriate mass matrix $\mathbf{Q}$ in the sense that there exist positive constants γ_1 and γ_2 such that

$$\gamma_1 \le \frac{\mathbf{q}^{\mathsf{T}}(\mathbf{N}+\mathbf{A}^{\mathsf{T}}\mathbf{M}^{-1}\mathbf{A})\mathbf{q}}{\mathbf{q}^{\mathsf{T}}\mathbf{Q}\mathbf{q}} \le \gamma_2$$

for all appropriate vectors $\mathbf{q}$ (Elman, Silvester, and Wathen, 2005). Similarly, it is typically assumed that

$$\gamma_1 \le \frac{\mathbf{v}^{\mathsf{T}}(\mathbf{M}+\mathbf{A}\mathbf{Q}^{-1}\mathbf{A}^{\mathsf{T}})\mathbf{v}}{\mathbf{v}^{\mathsf{T}}\mathbf{M}\mathbf{v}} \le \gamma_2$$

for all appropriate vectors $\mathbf{v}$. Under such assumptions, the preconditioner $\mathbf{H}$ defines a natural error norm, and it is spectrally equivalent to the error norm of Corollary 6.5.

Chapter 7

Software and Numerical Experiments

7.1 ▪ General Principle

One way to implement the methods of Chapter 5 is by modifying existing implementations of the standard LSQR and LSMR. Each code has two lines in its initialization that compute the first Golub–Kahan vectors $\bar{u}_1$ and $\bar{v}_1$. It suffices to replace those two lines with those of Algorithm 4.2 that compute u_1 and v_1 and the two lines in the main loop that compute $\bar{u}_{j+1}$ and $\bar{v}_{j+1}$ with the two lines in the main loop of Algorithm 4.2 using appropriate callbacks to solve systems with **M** and **N**.

7.2 ▪ Software Design

This book is accompanied by software implementations of the iterative methods in three programming languages. We like to say that all three are *high-productivity* languages. In their own way, each offers some form of object orientation, which is a key feature on which all three implementations rest:

1. MATLAB, which needs no introduction, is certainly one of the most widespread languages in computational science. In recent years, MATLAB has gained sufficiently mature and sophisticated object-oriented capabilities.

2. Python has become a mature challenger in the past decade. Python differs from MATLAB in that it was not designed as a language for computational science from the beginning. It is an elegant and powerful general-purpose object-oriented interpreted language that lends itself to beautiful and expressive constructs. It is particularly readable and easy to learn. Python has a number of traits of functional programming languages that often make it simple to express complicated ideas.

3. Finally, Julia is a recent addition to the computational scientist toolbelt, but one that holds much promise. The distinctive feature of Julia is that, though it is an interpreted language, functions are compiled just in time. What this means is that functions are automatically translated to low-level machine code and com-

piled when they are called for the first time. A second call to the function executes as a call to a compiled subprogram. The impact on run time can be drastic. Julia is still young and, at the time of this writing, has not yet reached version 1.0. However, it has already gained so much traction that we feel confident using it to develop scientific packages.

In order to understand the software design guidelines, it helps to remember that the iterative methods that we have described require only $\mathbf{A}$ as a proxy for the *linear operator* $\mathscr{A}$. In other words, they do not require access to $\mathbf{A}$ as an explicit matrix, although using an explicit matrix is certainly acceptable. In the same way, the inverses of $\mathbf{M}$ and $\mathbf{N}$ are required only as proxies for the linear operators that they represent. There is no need to explicitly invert matrices. As described in the introduction, if $\mathbf{M}$ represents a limited-memory BFGS approximation in an optimization context, it is never formed explicitly. Rather, we know efficient procedures that apply $\mathbf{M}$ to a vector. Even better, we know efficient procedures that apply $\mathbf{M}^{-1}$ to a vector!

In most iterative methods and, in particular, in Krylov methods, the most expensive operation at each iteration rests with applications of the operator and its adjoint. Our implementations assume that the user represented the operator as an abstract object that can be applied to vectors and whose adjoint is available and can also be applied to vectors. In all three languages, this assumption allows us to represent the linear operator $\mathscr{A}$ as an object, say **A**, such that, with reference to Figure 2.1, if $v \in \mathbb{N}$ is embodied by the variable `v`, a representation of the image $\mathscr{A}v$ can be computed by writing `A * v`. Similarly, if $u \in \mathbb{M}$ is represented by the variable `u`, a representation of the image $\mathscr{A}^\star u$ can be computed by writing `A' * u` in MATLAB and Julia and `A.T * u` in Python. The difference in notation results from "prime" being an operator in MATLAB and Julia and representing a transposition operation where it makes sense and that can be implemented for new objects. Python has no such operator. We instead implemented a method named **T** to represent the transpose of a linear operator. The net result is that general operators can be manipulated and applied as if they were explicit matrices, though access to elements is an expensive operation that requires applying the operator to a vector. This may appear simple, but there are a few advantages. First, the abstract notation is readable and helps the implementation look similar to the formal algorithm even in cases where $\mathbf{A}$ is not an explicit matrix. Second, representing operators as abstract objects allows us to stash distracting implementation details away from the user. A simple example where stashing details is useful is when we allocate memory beforehand to store the result of the application of $\mathbf{A}$ to a vector. Preallocating memory can yield large savings in run time and memory consumption, as computing `A * v` does not involve allocating a fresh memory area to store the result. Finally, we can query abstract operators essentially in the same way as matrices (e.g., an operator's size or data type—real, complex, or single or double precision), and operator objects can be used to store additional information in ways that functions cannot.

For the reasons described in the previous paragraph, the three implementations are based on a library of linear operators. All iterative methods take a linear operator and a right-hand side as input. It is up to the user to express the action of $\mathscr{A}$ as a linear operator object, but we provide shortcuts in all three languages. For instance, defining a linear operator based on a function $v \mapsto \mathscr{A}v$ is trivial. Similarly, the action of $\mathscr{M}^{-1}$ and $\mathscr{N}^{-1}$ must be defined as linear operators. A simple example consists of computing the Cholesky factorization of $\mathbf{M}$ and $\mathbf{N}$ and defining a linear operator object that retains those factors for future forward and backsolves. In optimization, $\mathbf{N}$

is often diagonal, and a linear operator that applies $\mathbf{N}^{-1}$ is easy to define.

7.3 ▪ Numerical Experiments

The numerical experiments of this section were performed with the Python implementation described in the PyKrylov (Orban, 2016) package.

We present numerical results on two categories of test cases: problems originating from optimization and from discretized partial-differential equations for fluid flow. In both cases the stopping conditions for all methods are those described in Algorithms 5.1, 5.2, 5.3, and 5.4 with $\tau = 10^{-12}$. This results in more iterations than would typically be necessary but illustrates the decrease of the relevant error estimates to tight tolerances. The stopping test used for DLANCZOS is that described in (6.11). The latter proved more reliable than a test based on even and odd iterations, as in practice, due to occasional instabilities in the factorization of the Lanczos tridiagonal, pivots do not always alternate sign at each iteration. As we explain below, it was necessary to select looser values of τ for this method. The stopping test used for MINRES is similar to those of Chapter 5 based on (6.27) with $\tau = 10^{-12}$.

Among other stopping conditions, both G-LSQR and G-LSMR declare optimality when either the relative residual or the relative residual of the normal equations falls below a certain threshold. More precisely, with $\epsilon_a = \epsilon_r = 10^{-12}$, the first stopping condition is

$$\frac{\bar{r}_k}{\|\mathbf{b}\|_2} \le \epsilon_a + \epsilon_r \frac{\|\mathbf{A}\|_2 \|\bar{\mathbf{y}}_k\|_2}{\|\mathbf{b}\|_2}, \tag{7.1}$$

where $\bar{r}_k$ is the value of the objective function of (2.4) at $\mathbf{y}_k = \mathbf{V}_k \bar{\mathbf{y}}_k$ and $\|\mathbf{A}\|_2$ is estimated by both G-LSQR and G-LSMR at each iteration. The second condition is

$$\frac{\bar{\rho}_k}{\|\mathbf{A}\|_2 \bar{r}_k} \le \epsilon_a, \tag{7.2}$$

where $\bar{\rho}_k$ is the right-hand side of (5.37). In our experiments, those stopping conditions were never the reason for terminating.

For the purpose of the numerical illustration below, linear systems involving $\mathbf{M}$ or $\mathbf{N}$ are solved using a one-time Cholesky factorization. In practice, the application should dictate the most appropriate solution method. It should be noted that no attempt was made to ensure clustering of the elliptic singular values of $\mathbf{A}$. Each problem name is followed by a tuple (n, m), where n is the order of $\mathbf{M}$ and m is the order of $\mathbf{N}$. For each iterative method, we report the history of the *relative* direct error in the appropriate metric together with *relative* error estimates for $d = 5$ and $d = 15$. At each iteration k, the direct error estimates are the quantities of the general form (5.11), while relative error estimates have the general form

$$\left(\sum_{j=k-d+1}^{k} \zeta_j^2 \Big/ \sum_{j=1}^{k} \zeta_j^2 \right)^{\frac{1}{2}}.$$

The error metric for each method is as given by Theorems 5.2, 5.4, 5.6, 5.8, and 6.2 and Corollary 6.5. Because DLANCZOS and MINRES combine two methods for the normal or Schur-complement equations, the value of d is doubled for them; i.e., setting $d = 5$ corresponds to an effective $d = 10$, so both underlying methods have time to converge.

Table 7.1 collects statistics on our test problems.

Table 7.1. *Summary of test problems.*

Name	Type	n	m	nnz(**M**)	nnz(**A**)	nnz(**N**)
DUAL1	Optimization	255	171	3728	425	171
STCQP1	Optimization	12291	10246	34797	29726	10246
COLLIDE	Stokes	578	289	2202	3465	1345
LID	Stokes	578	289	2202	3465	1345

7.3.1 ▪ Problems from optimization

The two problems below are generated at the third iteration of the primal-dual regularized interior-point method of Friedlander and Orban (2012). They originate from quadratic programming problems in standard form

$$\underset{\mathbf{x}\in\mathbb{R}^n}{\text{minimize}}\ \mathbf{g}^\mathsf{T}\mathbf{x}+\tfrac{1}{2}\mathbf{x}^\mathsf{T}\mathbf{H}\mathbf{x} \quad \text{subject to } \mathbf{C}\mathbf{x}=\mathbf{d},\ \mathbf{x}\geq 0,$$

where $\mathbf{g}\in\mathbb{R}^n$ and $\mathbf{H}=\mathbf{H}^\mathsf{T}\in\mathbb{R}^{n\times n}$ is positive semidefinite and result in linear systems with matrix

$$\begin{bmatrix}\mathbf{H}+\mathbf{X}^{-1}\mathbf{Z}+\rho I & \mathbf{C}^\mathsf{T}\\ \mathbf{C} & -\delta\mathbf{I}\end{bmatrix},$$

where $\rho>0$ and $\delta>0$ are regularization parameters. The system is initially shifted as in Section 5.1 to recover a right-hand side with all zeros in its last m components. The quadratic programs are part of the CUTEst collection (Gould, Orban, and Toint, 2015) and were chosen because they are representative of the behavior of the error curves. The numerical behavior of each method is illustrated in Figures 7.1 and 7.2.

Note that for both values of the window size d, the error estimates for G-LSQR and G-LSMR qualitatively follow the exact error, although they underestimate it by one or two orders of magnitude. This is typical of the problems we tested. In G-CRAIG and G-CRAIG-MR, the error estimates are not monotonic, and this behavior is more apparent on DUAL1. The exact error curve for both methods exhibits a temporary plateau, and at this point the error estimates try to recover from an under-estimation. Due to the window size, this recovery takes a number of iterations. The estimates otherwise closely follow the exact error curve. The hump in the error estimates is echoed in the G-MINRES curves, which combine G-LSMR and G-CRAIG-MR, located directly above it in the figure. Note that the hump occurs around iteration 80 for G-CRAIG-MR and around iteration 160 for MINRES, which confirms the fact that every other MINRES step is a G-CRAIG-MR step.

The curves for DLANCZOS are less intuitive. Note first that Theorem 6.2 states that the error is measured in an indefinite metric, while Theorem 6.1 explains that DLANCZOS steps alternate between G-LSQR and G-CRAIG steps, again located directly above the DLANCZOS plot in the figure. Following (6.10), the error changes between positive and negative values but globally decreases in absolute value. Figure 7.1 and those that follow plot the exact error curve for DLANCZOS on a *symmetric logarithmic scale*, i.e., a logarithmic scale in both positive and negative values. DLANCZOS turns out to be significantly less stable numerically than the other methods. Setting $\tau=10^{-12}$ results in failure for several problems due to roundoff errors, possibly including fatal loss of orthogonality and instability of the factorization of the Lanczos tridiagonal. Similarly, a window size of 15 iterations resulted in failures for at least one problem. Those failures are consistently due to the norm of the preconditioned residual becoming negative. Following this, DLANCZOS exits with an error message stating that the preconditioner is not positive definite. Another sign of numerical instability is that consecutive pivots occasionally have the same sign. We expect that an alternative implementation such as SYMMLQ (Paige and Saunders, 1975) would be more stable, although it is not yet clear how to recover the error estimates in SYMMLQ.

For the reasons above, the plots presented in this section for DLANCZOS use $\tau = 10^{-6}$ and window sizes of 5 and 10. In exact arithmetic, we expect that the oscillations of the exact error should be contained between two enveloping monotonic curves. It is almost the case for STCQP1, but we see that the curve exhibits a spike for DUAL1. It is not entirely clear what the origin of this spike is, but we speculate that it is partly due to DLANCZOS not exactly reducing to a decoupled combination of G-LSQR and G-CRAIG in finite-precision arithmetic. It also appears to overlap the plateau in the G-CRAIG error. The DLANCZOS error remains positive for a few iterations during which the G-CRAIG error does not decrease significantly. During those iterations, G-LSQR is essentially working alone toward reducing the error. For DUAL1, the error estimates are not particularly close to the exact error. For STCQP1, the error estimates follow the exact error more faithfully.

7.3.2 ▪ Problems from discretized PDEs

The two test cases of this section are discretizations of the stabilized Stokes equations for incompressible fluid flow (1.6) over a two-dimensional domain. The mesh and discretization are generated by the software IFISS 3.1 of Elman, Ramage, and Silvester (2007). The problems originate from Elman, Silvester, and Wathen (2005). In both problems, the domain is $\Omega := (-1,1) \times (-1,1)$, and the stabilization parameter is set to $\beta = 0.25$. These are examples where the regularization term $\mathbf{N}$ is not diagonal—in both examples, $\mathbf{N}$ is banded with semi-bandwidth 18. For both problems, the discretization occurs on a 16×16 grid with Q1-Q1 elements.

The first problem describes a colliding flow with analytic solution

$$u(x,y) = (20xy^3, 5x^4 - 5y^4), \quad p = 60x^2y - 20y^3 + \text{constant}$$

in Ω. Dirichlet boundary conditions are imposed along the whole boundary using the interpolant of the finite-element discretization of $u(x,y)$. Results are summarized in Figure 7.3.

The second problem describes a flow in a lid-driven regularized cavity. The lid velocity is given by $u(x,y) = (1 - x^4, 0)$. Results are summarized in Figures 7.3 and 7.4.

The behavior of the error estimates is smoother than for the optimization problems. In all cases, the error estimates for $d = 5$ and $d = 15$ are superposed and follow closely the exact error curve. For G-LSQR, G-CRAIG, G-LSMR, and G-CRAIG-MR, it would take about 30 iterations to reduce the error by a factor of 10^6.

We set the DLANCZOS stopping tolerance to $\tau = 10^{-3}$, which is reasonable given the discretization step size. For tighter tolerances, DLANCZOS again fails, complaining about an indefinite preconditioner, which indicates that the method has been driven past the numerical convergence point.

In the case of the lid-driven cavity problem, the exact error curve for MINRES increases in the last few iterations, and we believe that this is due to cancellation in the computation of the error. This behavior is visible to a lesser extent in G-LSQR and G-LSMR.

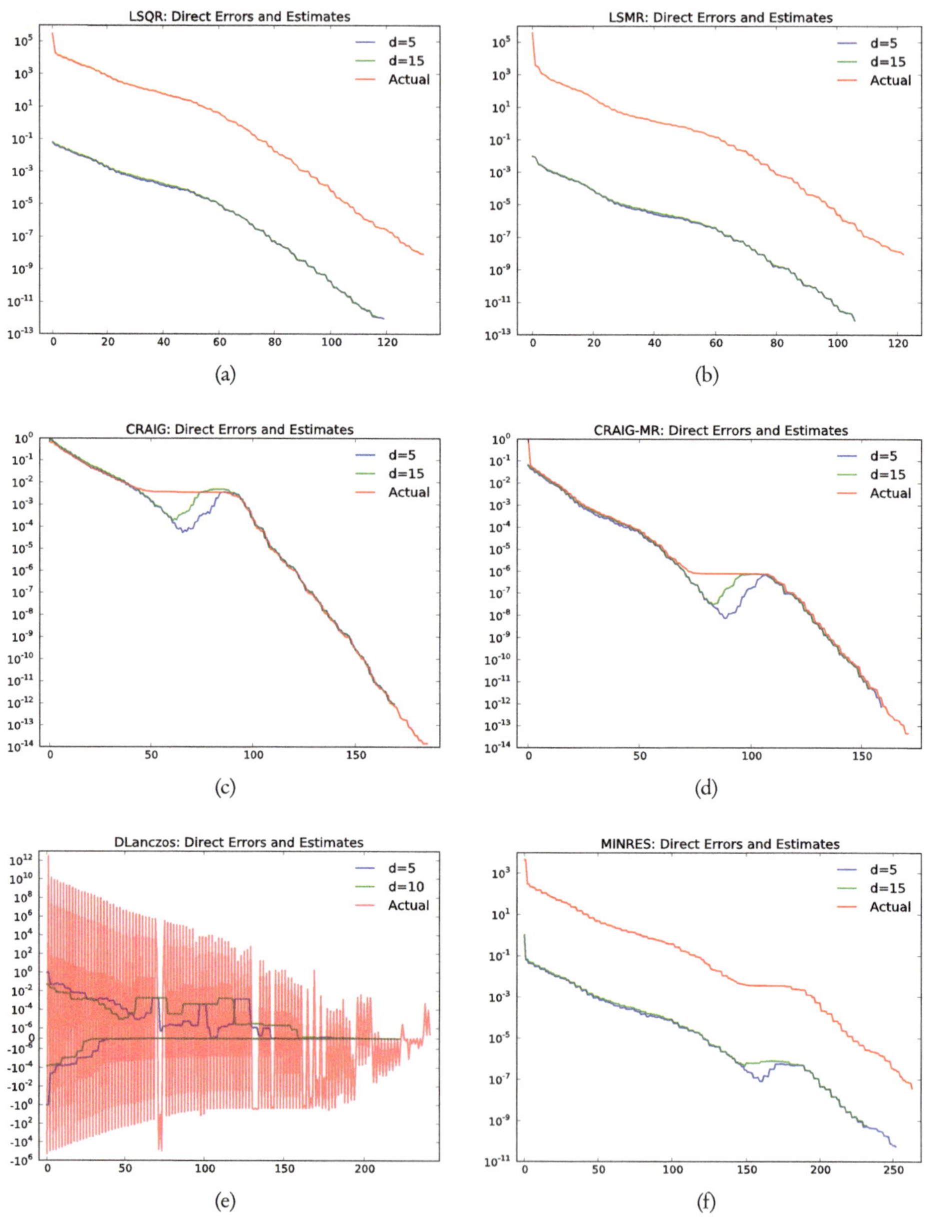

Figure 7.1. *Problem DUAL1* (255, 171). *Note the symmetric logarithmic scale of the vertical axis for* DLANCZOS *used to capture the fact that the error is measured in an indefinite metric.*

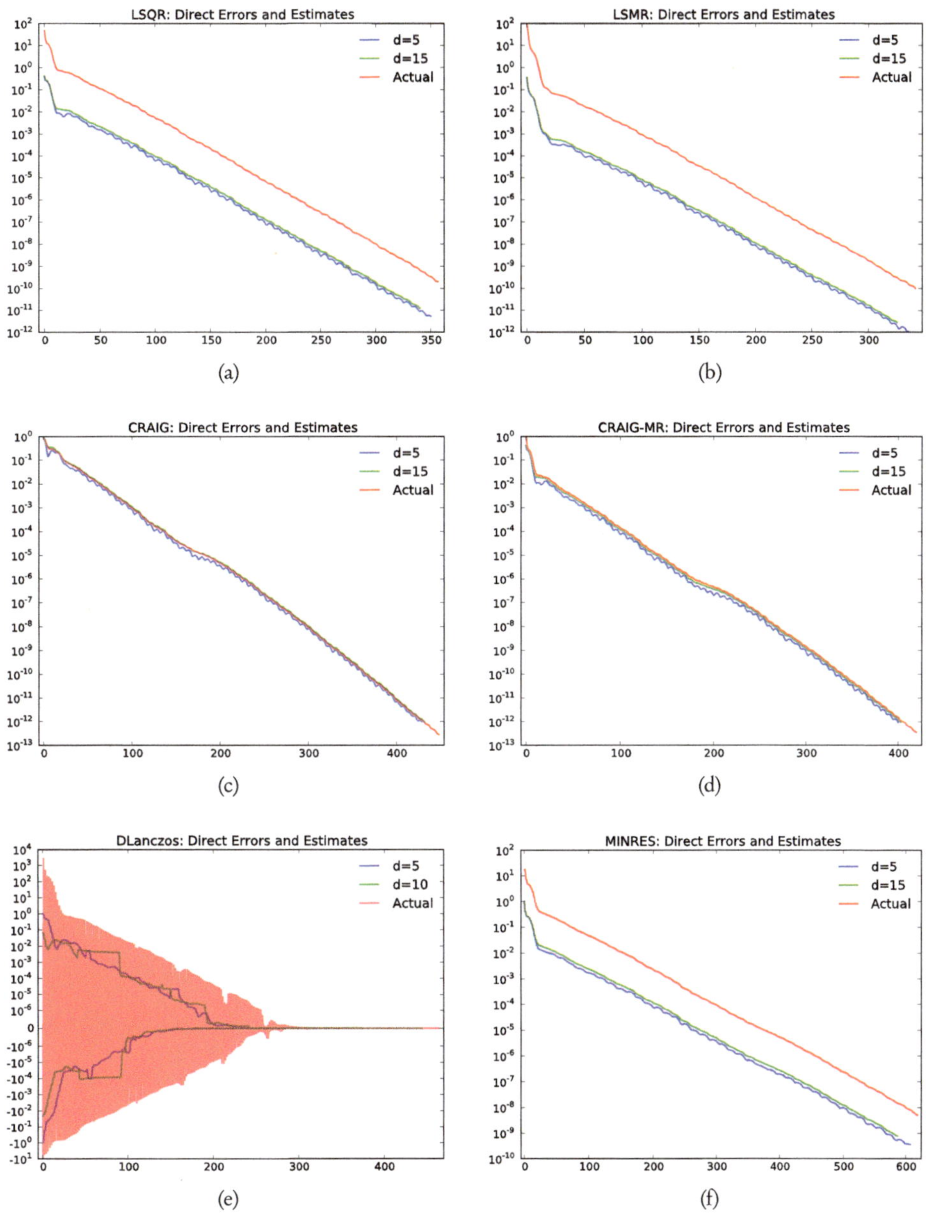

Figure 7.2. *Problem STCQP1* (12291, 10246). *Note the symmetric logarithmic scale of the vertical axis for* DLANCZOS *used to capture the fact that the error is measured in an indefinite metric.*

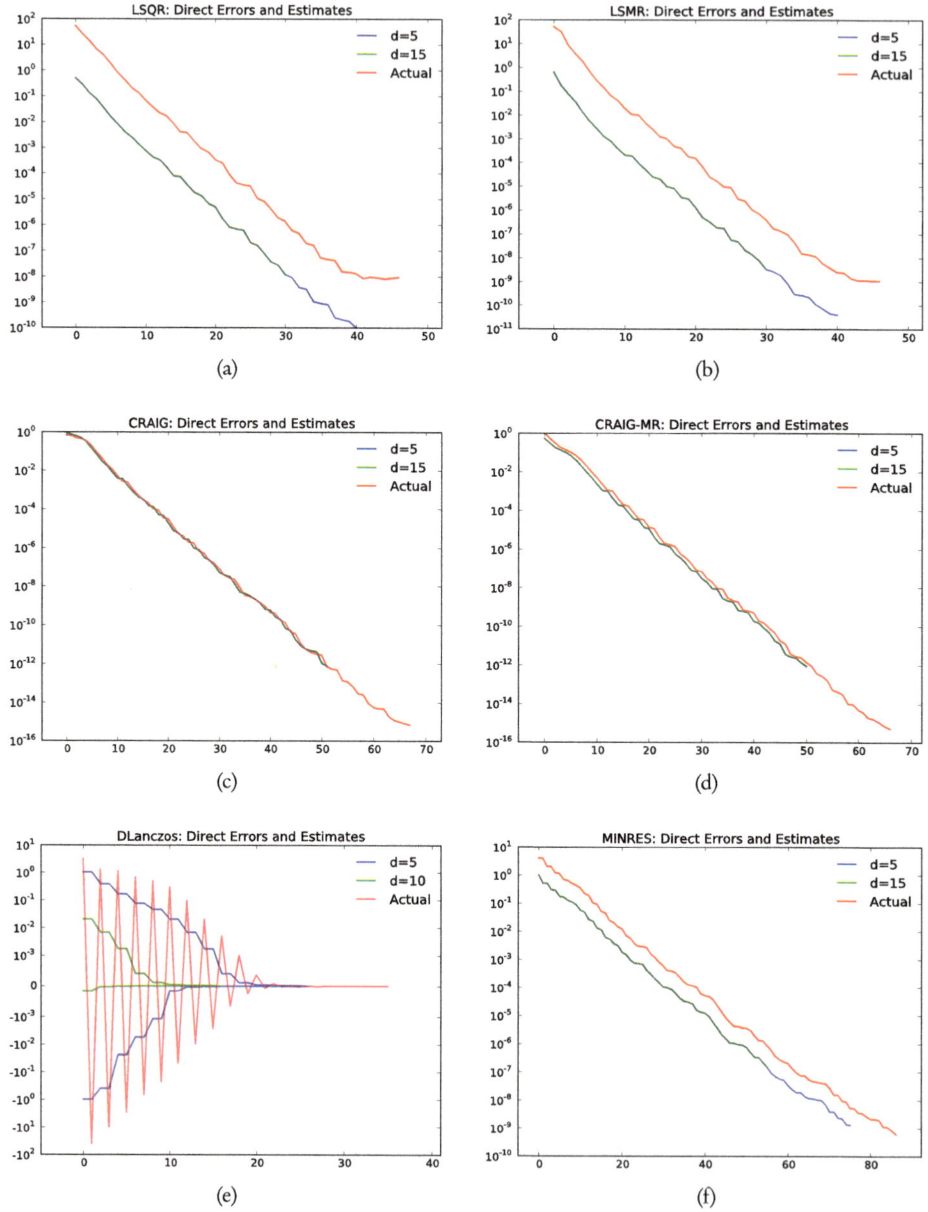

Figure 7.3. *Colliding flow* (578,289). *Note the symmetric logarithmic scale of the vertical axis for* DLANCZOS *used to capture the fact that the error is measured in an indefinite metric.*

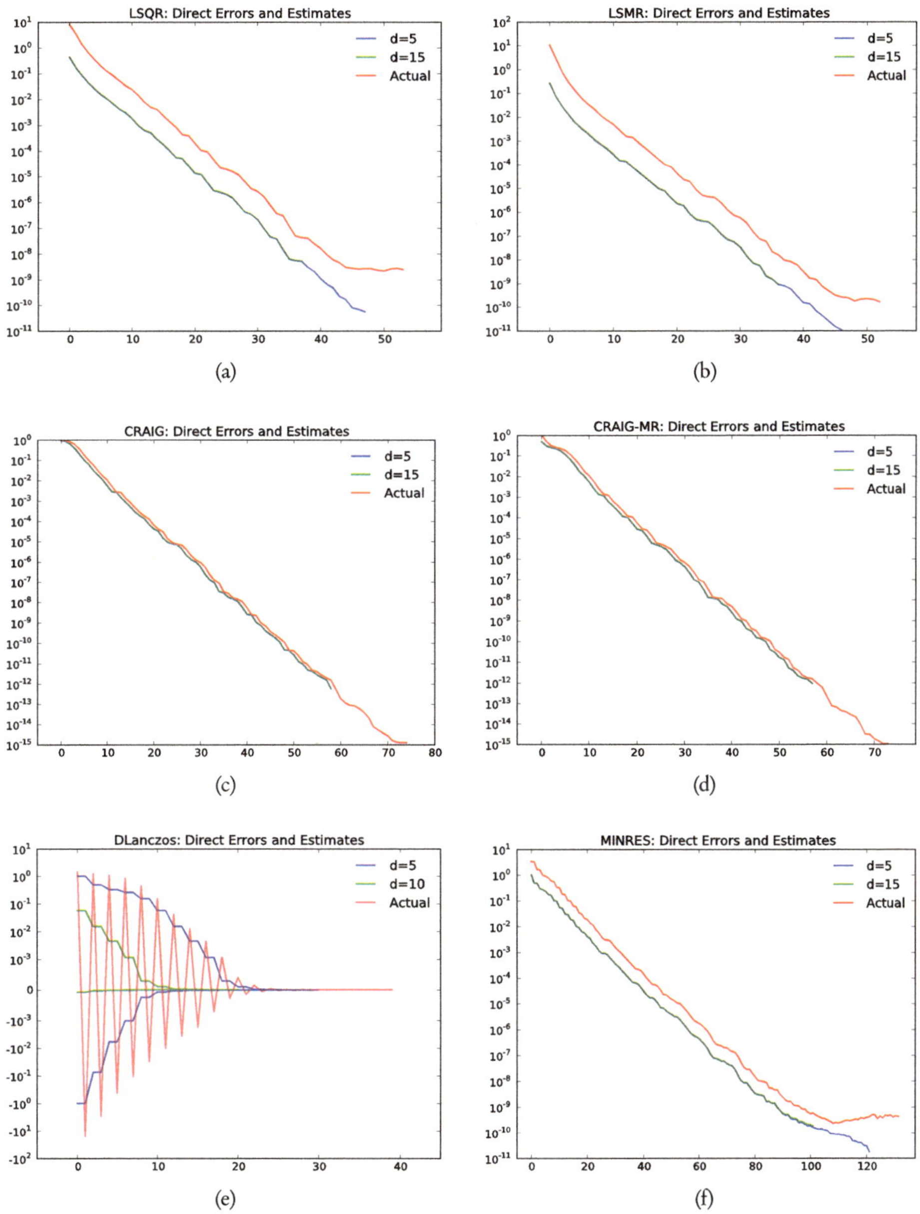

Figure 7.4. *Lid-driven cavity* (578, 289)*. Note the symmetric logarithmic scale of the vertical axis for* DLANCZOS *used to capture the fact that the error is measured in an indefinite metric.*

Chapter 8
Discussion and Open Questions

In this final chapter in our journey through iterative methods for SQD systems, we provide a few final comments and also point the interested reader in promising research directions.

The numerical experiments of Section 7.3 show that in all instances, methods for the normal or Schur-complement equations are attractive because the lower-bound estimates of the direct error follow the same trend as the exact error.

The behavior of DLANCZOS, and therefore CG and SYMMLQ, on SQD systems is instructive. Those methods solve the min-max problem

$$\underset{\mathbf{x}\in\mathbb{R}^n}{\text{minimize}}\ \underset{\mathbf{y}\in\mathbb{R}^m}{\text{maximize}}\ \mathscr{L}(\mathbf{x},\mathbf{y}),$$

where

$$\mathscr{L}(\mathbf{x},\mathbf{y}) := \tfrac{1}{2}\|\mathbf{x}\|_2^2 + \mathbf{x}^T\bar{\mathbf{A}}\mathbf{y} - \tfrac{1}{2}\|\mathbf{y}\|_2^2 - \mathbf{f}^T\mathbf{M}^{-\frac{1}{2}}\mathbf{x} - \mathbf{g}^T\mathbf{N}^{-\frac{1}{2}}\mathbf{y},$$

where $\bar{\mathbf{A}} = \mathbf{M}^{-\frac{1}{2}}\mathbf{A}\mathbf{N}^{-\frac{1}{2}}$, and whose first-order optimality conditions coincide with (1.5) preconditioned with $\mathbf{H}$ defined in (2.12), and in which the sign of the error alternates. Every other step is a minimization step on the convex part of $\mathscr{L}$, i.e., the function $\mathbf{x} \mapsto \mathscr{L}(\mathbf{x},\mathbf{y})$ for fixed $\mathbf{y}$, while the next step is a maximization step on the concave part of $\mathscr{L}$, i.e., $\mathbf{y} \mapsto \mathscr{L}(\mathbf{x},\mathbf{y})$ for fixed $\mathbf{x}$.

Because the spectrum of (1.5) consists of a set of negative and a set of positive eigenvalues, two observations are in order:

1. If we wish to develop upper-bound estimates of the forward error, we must abandon the positive measure underpinning the Gauss–Radau strategy used in Section 5.9—instead, it may be possible to develop a similar approach based on general nonpositive Radon measures.

2. The form of the spectrum and the fact that it is symmetric suggest that it could be interesting to use some of the ideas described by Anderssen and Golub (1972) and to develop a generalized Chebyshev approach—see Fischer (2011, Section 3.3) for an introduction to this topic.

We emphasize that we have concentrated on the case where systems with the diagonal blocks $\mathbf{M}$ and $\mathbf{N}$ are easily solved. There are clearly numerous practical cases where this assumption is not realistic. For instance, in optimization applications, $\mathbf{M}$

may represent a (possibly dense) quadratic term that is not easily inverted. It remains relatively typical that systems with **N** are easily solved. Again, in optimization, **N** is usually diagonal. The same is not necessarily true in fluid flow applications where **N** may represent a mass matrix, but Wathen (1987) shows that such systems are efficiently solved in a few iterations of the conjugate-gradient method with diagonal preconditioner. The study of the general case, possibly allowing for inexact solves with **M** and/or **N**, is an interesting research direction. Certainly, this includes, as a special case, a finite-precision arithmetic extension of our framework. More generally, one might imagine computing only incomplete Cholesky factorizations of **M** and/or **N**. How accurately must we solve systems with **M** and **N** in order to retain convergence?

The study of symmetric quasi-definite systems in the context of preconditioning is also an interesting research candidate. A first important aspect is the preconditioning of SQD systems. In particular, not all preconditioners preserve the SQD structure. It is clear that if **P** and **Q** are symmetric and positive definite, the preconditioner blkdiag(**P**, **Q**) preserves the SQD structure. If the more general symmetric preconditioner

$$\begin{bmatrix} \mathbf{P} & \mathbf{R} \\ \mathbf{R}^T & \mathbf{Q} \end{bmatrix}$$

applied to the left- and right-hand sides of (1.5) is to yield and SQD system, it must satisfy

$$\mathbf{PMP} + \mathbf{RA}^T\mathbf{P} + \mathbf{PAR}^T - \mathbf{RNR}^T \succ 0,$$
$$\mathbf{R}^T\mathbf{MR} + \mathbf{QA}^T\mathbf{R} + \mathbf{R}^T\mathbf{AQ} - \mathbf{QNQ} \prec 0.$$

To the best of our knowledge, no preconditioning strategy has yet been designed based on the above conditions. We note, however, that Orban (2015) develops a limited-memory factorization of SQD matrices based on their strong factorizability property. The factorization yields a unit lower triangular **L** and an indefinite diagonal **D**. Those factors may be used to form an indefinite preconditioner $\mathbf{LDL}^T$ for use with, e.g., GMRES (Saad and Schultz, 1986), or a positive-definite preconditioner $\mathbf{L}|\mathbf{D}|\mathbf{L}^T$ for use with MINRES.

A second important aspect is that SQD operators may be used to precondition standard saddle-point systems, whether symmetric or not. For instance, systems encountered during an iterative process to solve the Navier–Stokes equations typically have a zero (2,2) block. As pointed out by Benzi, Golub, and Liesen (2005), it is possible to devise efficient SQD preconditioners for such systems.

The methods presented in this book are relevant to optimization contexts beyond the occurrence of SQD systems. Indeed, in trust-region-based Gauss–Newton methods, subproblems such as (2.4) must be solved at each iteration, but an accurate solution is not necessarily sought. Rather, sufficient decrease in the residual is acceptable, and a constraint on the maximal norm of **y** is imposed. It is a property of standard LSQR and LSMR that the iterates $\bar{y}_k$ increase in Euclidean norm (Paige and Saunders, 1982; Fong and Saunders, 2011). Thus, the quantities $\|\mathbf{y}_k\|_\mathbf{N}$ increase along the generalized LSQR and LSMR iterations. Consequently, it is reasonable to solve (2.4) with a trust-region constraint of the form $\|\mathbf{y}\|_\mathbf{N} \leq \Delta$ for some trust-region radius $\Delta > 0$ using the initial guess $\mathbf{y}_0 = 0$. As LSQR is equivalent to the conjugate-gradient method on the normal equations, interrupting the iterations as soon as the boundary of the trust-region is crossed ensures sufficient decrease. In the case of LSMR, it remains necessary to establish that the decrease thus obtained is a fraction of that obtained at the Cauchy point—we refer the interested reader to Conn, Gould, and Toint (2000, Chapter 7) for details.

All methods covered above apply equally to the SQD system (1.5) with $\mathbf{f} = 0$ and $\mathbf{g} = \mathbf{b}$. Indeed, the system can be reduced to the normal equations

$$(\mathbf{A}\mathbf{N}^{-1}\mathbf{A}^T + \mathbf{M})\mathbf{x} = \mathbf{A}\mathbf{N}^{-1}\mathbf{b},$$

which are the optimality conditions of the regularized and weighted linear least-squares problem

$$\underset{\mathbf{x}\in\mathbb{R}^n}{\text{minimize}}\ \frac{1}{2}\left\| \begin{bmatrix} \mathbf{A}^T \\ \mathbf{M}^{\frac{1}{2}} \end{bmatrix} \mathbf{x} - \begin{bmatrix} \mathbf{b} \\ 0 \end{bmatrix} \right\|^2_{\mathbf{N}_+^{-1}}.$$

All methods presented in this book are based on the lower bidiagonalization procedure referred to as "bidiag1" by Paige and Saunders (1982). A corresponding family of numerical methods may also be derived from the "bidiag2" procedure in the same paper, which performs an upper bidiagonalization of $\mathbf{A}$ and is initialized with $\mathbf{A}^T\mathbf{b}$ instead of $\mathbf{b}$. Two variants of the Golub–Kahan process based on inner products defined by $\mathbf{M}$ and $\mathbf{N}$ and corresponding to Algorithms 4.2 and 4.3 give rise to alternative generalizations of LSQR, CRAIG, CG, LSMR, CRAIG-MR, and MINRES. Whether one of those methods dominates the others numerically should be determined via intensive testing. Arioli (2013) derives a generalized variant of CRAIG based on "bidiag2" in the case where the bottom block of the matrix in (1.5) is zero yet there exists an appropriate metric $\mathbf{N}$ to measure the norm of $\mathbf{y}$.

It does not appear possible to apply the conjugate gradient to SQD systems in general if the right-hand side does not have the form $(\mathbf{b}, 0)$. Consider, for instance, the SQD system

$$\begin{bmatrix} \mathbf{I} & \\ & -\mathbf{I} \end{bmatrix} \begin{bmatrix} \mathbf{x} \\ \mathbf{y} \end{bmatrix} = \begin{bmatrix} \mathbf{e} \\ \mathbf{e} \end{bmatrix},$$

where $\mathbf{e}$ is the vector of all ones. At the first iteration, the conjugate-gradient methods needs to generate the denominator

$$\begin{bmatrix} \mathbf{e}^T & \mathbf{e}^T \end{bmatrix} \begin{bmatrix} \mathbf{I} & \\ & -\mathbf{I} \end{bmatrix} \begin{bmatrix} \mathbf{e} \\ \mathbf{e} \end{bmatrix} = 0$$

and must break down.

Remember that all our analysis is made under the assumption of exact arithmetic. In finite precision arithmetic, such as the IEEE 64-bit standard, we can expect our algorithms to behave differently. Owing to the analogies between them and the conjugate-gradient method and/or the Lanczos algorithm, we are expecting a behavior similar to that analyzed by Greenbaum (1989), Greenbaum and Strakoš (1992), and Meurant (2006): delay in convergence, i.e., slower descent of the error, and a potential instability in the computation of the upper bounds of the errors (Golub and Strakoš, 1994; Strakoš and Tichy, 2002) provide an analysis of the bounds relative to the conjugate gradient. Our numerical experiments support good stability properties of the lower-bound estimates as Strakoš and Tichy (2002) assert for the classical conjugate gradient. In particular, when the SQD system arises from an approximation of differential problems, the bounds on the error are reliable. However, users must be aware of the potential instabilities—see the numerical experiments discussion by Arioli (2004).

We gave an interpretation of the conjugate-gradient method applied to a SQD system with appropriate right-hand side in terms of a min-max problem on a saddle-point function and in terms of a combination of LSQR and CRAIG. MINRES performs twice as much work as is really necessary because it aims to minimize the residual of both the normal and Schur-complement equations. Computations can be saved by employing only G-LSMR or G-CRAIG-MR.

It appears from our analysis that, on the one hand, the generalized LSQR and CRAIG and, on the other hand, the generalized LSMR and CRAIG-MR are the appropriate implementations of CG and MINRES for SQD systems.

Bibliography

R. S. Anderssen and G. H. Golub. *Richardson's Non-stationary Matrix Iterative Procedure*. Technical Report STAN-CS-72-304, Stanford University, 1972. (Cited on p. 83)

M. Arioli. A stopping criterion for the conjugate gradient algorithm in a finite element method framework. *Numerische Mathematik*, 97(1):1–24, 2004. DOI: 10.1007/s00211-003-0500-y. (Cited on p. 85)

M. Arioli. Generalized Golub–Kahan bidiagonalization and stopping criteria. *SIAM Journal on Matrix Analysis and Applications*, 34(2):571–592, 2013. DOI: 10.1137/120866543. (Cited on pp. 8, 32, 39, 43, 46, 59, 85)

M. Arioli and I. S. Duff. Using FGMRES to obtain backward stability in mixed precision. *Electronic Transactions on Numerical Analysis*, 33:31–44, 2008/2009. http://etna.mcs.kent.edu/vol.33.2008-2009/pp31-44.dir/pp31-44.pdf. (Cited on pp. 22, 23)

M. Arioli, I. S. Duff, S. Gratton, and S. Pralet. A note on GMRES preconditioned by a perturbed LDL^T decomposition with static pivoting. *SIAM Journal on Scientific Computing*, 29(5):2024–2044, 2007. DOI: 10.1137/060661545. (Cited on pp. 22, 23)

M. Arioli and I. S. Duff. Preconditioning linear least-squares problems by identifying a basis matrix. *SIAM Journal on Scientific Computing*, 37(5):S544–S561, 2015. DOI: 10.1137/140975358. (Cited on p. 26)

O. Axelsson. *Iterative Solution Methods*. Cambridge University Press, Cambridge, UK, 1994. (Cited on p. 25)

O. Axelsson and M. Neytcheva. Preconditioning methods for linear systems arising in constrained optimization problems. *Numerical Linear Algebra with Applications*, 10(1–2):3–31, 2003. (Cited on p. 7)

S. J. Benbow. Solving generalized least-squares problems with LSQR. *SIAM Journal on Matrix Analysis and Applications*, 21(1):166–177, 1999. DOI: 10.1137/S0895479897321830. (Cited on pp. 8, 31, 32, 36)

M. Benzi, G. H. Golub, and J. Liesen. Numerical solution of saddle point problems. *Acta Numerica*, 14:1–137, 2005. DOI: 10.1017/S0962492904000212. (Cited on pp. 1, 7, 8, 84)

Å. Björck. *Numerical Methods for Least Squares Problems*. SIAM, Philadelphia, PA, 1996. DOI: 10.1137/1.9781611971484. (Cited on p. 3)

H. Brézis. *Analyse fonctionnelle: Théorie et applications*. Dunod, Paris, 1983. (Cited on pp. 11, 12)

A. Bunse-Gerstner. Private communication, 2012. (Cited on p. 6)

R. Chandra. *Conjugate Gradient Methods for Partial Differential Equations*. Ph.D. Thesis, Department of Computer Science, Yale University, New Haven, CT, 1978. (Cited on p. 7)

P. Concus and G. H. Golub. A generalized conjugate gradient method for non-symmetric systems of linear equations. In R. Glowinski and J. L. Lions, editors, *Second International Symposium on Computing Methods in Applied Sciences and Engineering: Summaries: Versailles, December, 15–19, 1975*, Lecture Notes in Economics and Mathematical Systems 134, Springer-Verlag, New York, 1975, pp. 56–65. (Cited on p. 26)

A. R. Conn, N. I. M. Gould, and Ph. L. Toint. *Trust-Region Methods*. SIAM, Philadelphia, PA, 2000. (Cited on p. 85)

P. Courtier. Dual formulation of four-dimensional variational assimilation. *Quarterly Journal of the Royal Meteorological Society*, 123:2449–2461, 1997. DOI: 10.1002/qj.49712354414. (Cited on p. 6)

J. E. Craig. The N-step iteration procedures. *Journal of Mathematics and Physics*, 34(1): 64–73, 1955. (Cited on pp. 3, 5, 11, 41)

H. S. Dollar, N. I. M. Gould, W. H. A. Schilders, and A. J. Wathen. Implicit-factorization preconditioning and iterative solvers for regularized saddle-point systems. *SIAM Journal on Matrix Analysis and Applications*, 28(1):170–189, 2006. DOI: 10.1137/05063427X. (Cited on pp. 7, 24)

I. S. Duff. MA57—a code for the solution of sparse symmetric definite and indefinite systems. *ACM Transactions on Mathematical Software*, 30(2):118–144, 2004. DOI: 10.1145/992200.992202. (Cited on pp. 5, 21, 23)

I. S. Duff, A. M. Erisman, and J. K. Reid. *Direct Methods for Sparse Matrices*. Numerical Mathematics and Scientific Computation, Oxford University Press, Oxford, UK, second edition, 2017. ISBN 9780198508380. (Cited on p. 21)

H. Elman, D. Silvester, and A. Wathen. *Finite Elements and Fast Iterative Solvers with Applications in Incompressible Fluid Dynamics*. Numerical Mathematics and Scientific Computation, Oxford University Press, Oxford, UK, 2005. (Cited on pp. 72, 77)

H. C. Elman, A. Ramage, and D. J. Silvester. Algorithm 866: IFISS, a Matlab toolbox for modelling incompressible flow. *ACM Transactions on Mathematical Software*, 33: article 14, 2007. DOI: 10.1145/1236463.1236469. (Cited on p. 77)

B. Fischer. *Polynomial Based Iteration Methods for Symmetric Linear Systems*. Classics in Applied Mathematics 68. SIAM, Philadelphia, PA, 2011. DOI: 10.1137/1.9781611971927. Originally published 1996. (Cited on pp. 14, 15, 83)

B. Fischer, A. Ramage, D. J. Silvester, and A. J. Wathen. Minimum residual methods for augmented systems. *BIT*, 38(3):527–543, 1998. (Cited on p. 4)

D. C.-L. Fong and M. A. Saunders. LSMR: An iterative algorithm for sparse least-squares problems. *SIAM Journal on Scientific Computing*, 33(5):2950–2971, 2011. DOI: 10.1137/10079687X. (Cited on pp. 3, 5, 35, 46, 48, 51, 53, 84)

R. W. Freund, G. H. Golub, and N. M. Nachtigal. Iterative solutions of linear systems. *Acta Numerica*, 1:1–44, 1991. DOI: 10.1017/S0962492900002245. (Cited on p. 15)

M. P. Friedlander and D. Orban. A primal-dual regularized interior-point method for convex quadratic programs. *Mathematical Programming Computation*, 4(1):71–107, 2012. DOI: 10.1007/s12532-012-0035-2. (Cited on pp. 10, 76)

A. George and Kh. Ikramov. On the condition of symmetric quasi-definite matrices. *SIAM Journal on Matrix Analysis and Applications*, 21(3):970–977, 2000. DOI: 10.1137/S0895479898333247. (Cited on pp. 8, 21)

A. George, Kh. Ikramov, and A. B. Kucherov. Some properties of symmetric quasi-definite matrices. *SIAM Journal on Matrix Analysis and Applications*, 21:1318–1323, 2000. DOI: 10.1137/S0895479897329400. (Cited on p. 8)

P. E. Gill, M. A. Saunders, and J. R. Shinnerl. On the stability of Cholesky factorization for symmetric quasidefinite systems. *SIAM Journal on Matrix Analysis and Applications*, 17(1):35–46, 1996. DOI: 10.1137/S0895479893252623. (Cited on pp. 8, 19, 20, 65)

I. Gohberg, P. Lancaster, and L. Rodman. *Indefinite Linear Algebra and Applications*. Birkhäuser, Basel, Switzerland, 2005. (Cited on p. 67)

G. H. Golub and W. Kahan. Calculating the singular values and pseudo-inverse of a matrix. *SIAM Journal on Numerical Analysis*, 2(2):205–224, 1965. DOI: 10.1137/0702016. (Cited on pp. 2, 4, 5, 30)

G. H. Golub and G. Meurant. *Matrices, Moments and Quadrature with Applications*. Princeton University Press, Princeton, NJ, 2010. (Cited on pp. 39, 55, 57, 58, 59, 60)

G. H. Golub and C. Van Loan. Unsymmetric positive definite linear systems. *Linear Algebra and Its Applications*, 28:85–97, 1979. ISSN 0024-3795. DOI: 10.1016/0024-3795(79)90122-8. (Cited on pp. 20, 21)

G. H. Golub and C. F. Van Loan. *Matrix Computations*. Johns Hopkins University Press, Baltimore, MD, fourth edition, 2013. ISBN 978-1-4214-0794-4. (Cited on pp. 20, 21)

G. H. Golub and G. Meurant. Matrices, moments and quadrature II: How to compute the norm of the error in iterative methods. *BIT*, 37(3):687–705, 1997. DOI: 10.1007/BF02510247. (Cited on pp. 39, 57)

G. H. Golub and Z. Strakoš. Estimates in quadratic formulas. *Numerical Algorithms*, 8:241–268, 1994. (Cited on p. 85)

N. I. M. Gould, D. Orban, and T. Rees. *Projected Krylov Methods for Saddle-Point Systems*. Cahier du GERAD G-2013-23, GERAD, Montréal, QC, Canada, 2013. (Cited on p. 24)

N. I. M. Gould, D. Orban, and Ph. L. Toint. CUTEst: A constrained and unconstrained testing environment with safe threads for mathematical optimization. *Computational Optimization and Applications*, 60:545–557, 2015. DOI: 10.1007/s10589-014-9687-3. (Cited on p. 76)

A. Greenbaum. Behavior of slightly perturbed Lanczos and conjugate-gradient recurrences. *Linear Algebra and Applications*, 113:7–63, 1989. (Cited on p. 85)

A. Greenbaum and Z. Strakoš. Predicting the behavior of finite precision Lanczos and conjugate gradient computations. *SIAM Journal on Matrix Analysis and Applications*, 13(1):121–137, 1992. (Cited on p. 85)

Harwell Subroutine Library. *A Collection of Fortran Codes for Large-Scale Scientific Computation*. AERE Harwell Laboratory, http://www.hsl.rl.ac.uk, 2000. (Cited on p. 21)

M. R. Hestenes and E. Stiefel. Methods of conjugate gradients for solving linear systems. *Journal of Research of the National Bureau of Standards*, 49(6):409–436, 1952. (Cited on pp. 2, 5, 39)

R. E. Kalman. A new approach to linear filtering and prediction problems. *Transactions of the ASME—Journal of Basic Engineering*, 82(Series D):35–45, 1960. (Cited on p. 6)

C. Keller, N. I. M. Gould, and A. J. Wathen. Constraint preconditioning for indefinite linear systems. *SIAM Journal on Matrix Analysis and Applications*, 21(4):1300–1317, 2000. DOI: 10.1137/S0895479899351805. (Cited on p. 24)

J. Korzak. Eigenvalue relations and conditions of matrices arising in linear programming. *Computing*, 62(1):45–54, 1999. DOI: 10.1007/s006070050012. (Cited on p. 8)

C. Lanczos. An iteration method for the solution of the eigenvalue problem of linear differential and integral operators. *Journal of Research of the National Bureau of Standards*, 45(4):255–282, 1950. (Cited on p. 29)

C. Lanczos. Solution of systems of linear equations by minimized iterations. *Journal of Research of the National Bureau of Standards*, 49(1):33–53, 1952. (Cited on pp. 2, 29)

C.-J. Lin and J. J. Moré. Incomplete Cholesky factorizations with limited memory. *SIAM Journal on Scientific Computing*, 21(1):24–45, 1999. DOI: 10.1137/S1064827597327334. (Cited on p. 25)

R. F. Marcia. On solving sparse symmetric linear systems whose definiteness is unknown. *Applied Numerical Mathematics*, 58(4):449–458, 2008. DOI: 10.1016/j.apnum.2007.01.014. (Cited on pp. 8, 65)

G. Meurant. *The Lanczos and Conjugate Gradient Algorithms, from Theory to Finite Precision Computations*. SIAM, Philadelphia, PA, 2006. (Cited on p. 85)

J. Nocedal and S. J. Wright. *Numerical Optimization*. Springer Series in Operations Research, Springer-Verlag, New York, second edition, 2006. (Cited on p. 6)

D. Orban. Limited-memory LDL^T factorization of symmetric quasi-definite matrices with application to constrained optimization. *Numerical Algorithms*, 70(1):9–41, 2015. DOI: 10.1007/s11075-014-9933-x. (Cited on pp. 25, 84)

D. Orban. PyKrylov: Krylov subspace methods in pure Python. https://github.com/PythonOptimizers/pykrylov, February 2016. (Cited on p. 75)

C. C. Paige. Bidiagonalization of matrices and solution of linear equations. *SIAM Journal on Numerical Analysis*, 11(1):197–209, 1974. DOI: 10.1137/0711019. (Cited on pp. xiv, 31, 41)

C. C. Paige and M. A. Saunders. Solution of sparse indefinite systems of linear equations. *SIAM Journal on Numerical Analysis*, 12(4):617–629, 1975. DOI: 10.1137/0712047. (Cited on pp. 2, 5, 7, 46, 70, 71, 77)

C. C. Paige and M. A. Saunders. LSQR: An algorithm for sparse linear equations and sparse least squares. *ACM Transactions on Mathematical Software*, 8(1):43–71, 1982. DOI: 10.1145/355984.355989. (Cited on pp. 3, 5, 11, 35, 38, 39, 84, 85)

I. Perugia and V. Simoncini. Block-diagonal and indefinite symmetric preconditioners for mixed finite element formulations. *Numerical Linear Algebra with Applications*, 7:585–616, 2000. (Cited on p. 7)

Y. Saad. *Iterative Methods for Sparse Linear Systems*. SIAM, Philadelphia, PA, second edition, 2003. DOI: 10.1137/1.9780898718003. (Cited on p. 64)

Y. Saad and M. H. Schultz. GMRES: A generalized minimal residual algorithm for solving nonsymmetric linear systems. *SIAM Journal on Scientific and Statistical Computing*, 7(3):856–869, 1986. DOI: 10.1137/0907058. (Cited on p. 84)

Y. Saad. A flexible inner-outer preconditioned GMRES algorithm. *SIAM Journal on Scientific and Statistical Computing*, 14(2):461–469, 1993. (Cited on p. 23)

M. A. Saunders. Solution of sparse rectangular systems using LSQR and CRAIG. *BIT*, 35:588–604, 1995. DOI: 10.1007/BF01739829. (Cited on pp. xiv, 8, 14, 36, 40, 41, 43)

J. Scott and M. Tůma. On signed incomplete Cholesky factorization preconditioners for saddle-point systems. *SIAM Journal on Scientific Computing*, 36(6):A2984–A3010, 2014. DOI: 10.1137/140956671. (Cited on p. 25)

D. Silvester and A. Wathen. Fast iterative solution of stabilised Stokes systems part II: Using general block preconditioners. *SIAM Journal on Numerical Analysis*, 31(5):1352–1367, 1994. DOI: 10.1137/0731070. (Cited on p. 6)

D. J. Silvester and V. Simoncini. An optimal iterative solver for symmetric indefinite systems stemming from mixed approximation. *ACM Transactions on Mathematical Software*, 37(4):42:1–42:22, 2011. DOI: 10.1145/1916461.1916466. (Cited on p. 72)

Z. Strakoš and P. Tichy. On error estimation in the conjugate gradient method and why it works in finite precision computations. *Electronic Transactions on Numerical Analysis*, 13:56–80, 2002. (Cited on p. 85)

G. Strang. *Introduction to Applied Mathematics*. Wellesley-Cambridge Press, Wellesley, MA, 1986. (Cited on p. 6)

D. B. Szyld and O. B. Widlund. Variational analysis of some conjugate gradient methods. *East-West Journal of Numerical Mathematics*, 1:51–74, 1993. (Cited on pp. 26, 27)

H. A. van der Vorst. *Iterative Krylov Methods for Large Linear Systems*, Cambridge Monographs on Applied and Computational Mathematics 13, Cambridge University Press, Cambridge, UK, 2009. Paperback edition of the 2003 original text. (Cited on p. 33)

R. J. Vanderbei. Symmetric quasidefinite matrices. *SIAM Journal on Optimization*, 5 (1):100–113, 1995. DOI: 10.1137/0805005. (Cited on pp. 4, 5, 17, 18, 19)

A. Wathen. Realistic eigenvalue bounds for the Galerkin mass matrix. *IMA Journal of Numerical Analysis*, 7(4):449–457, 1987. DOI: 10.1093/imanum/7.4.449. (Cited on pp. 11, 84)

D. Watkins. *The Matrix Eigenvalue Problem: GR and Krylov Subspace Methods*. SIAM, Philadelphia, PA, 2007. DOI: 10.1137/1.9780898717808. (Cited on p. 33)

O. Widlund. A Lanczos method for a class of nonsymmetric systems of linear equations. *SIAM Journal on Numerical Analysis*, 15(4):801–812, 1978. (Cited on pp. 26, 27)

F. Zhang. *The Schur Complement and Its Applications*, Numerical Methods and Algorithms 4, Springer, New York, 2005. DOI: 10.1007/b105056. (Cited on pp. 17, 18)

Index